"Sustainability goes above and beyond a set of actionable goals – it should constitute a conducting thread that permeates the mission, vision and values of our communities, organizations and each one of us. This very timely and uplifting collection of stories inspires educators and practitioners to instill and nurture a sustainability mindset in those who would lead future generations on the transformative journey."

**Eric Cornuel**, *President, EFMD (Belgium)*

"This book is a must read for anyone with an interest in understanding how to engage students, educators and leaders but also 'our own selves' on a personal journey to support future betterment of our common world. The engagement of students, educators and leaders as authors in an achievement in itself. Congratulations!"

**Mette Morsing**, *Global Head, Principles for Responsible Management Education, United Nations Global Compact (USA)*

"Life is a change. Living now means continuously adjusting to accelerating speed of change. The grand challenge for education is to develop open minds and creativity among students, and together with them paving the roads to a sustainable future. This book, however, is not about how professional educators see opportunities to help students discover these roads. It offers stories of more than 150 students, the generation to make the next steps, about how transformational experience shifted their mindsets. For educators, mentors and consultants, this is a wealth of information and tools to be able to deliver an excellent and relevant professional contribution in helping to build a sustainable future and a better world."

**Prof. Danica Purg**, *President of IEDC-Bled School of Management, President of CEEMAN (Slovenia)*

"The right mindset is instrumental to drive the sustainability transformation. This book provides insights about how to change the mindset and how to accomplish the transformation. A must read!"

**Katrin Muff**, *Director, the Institute for Business Sustainability (Switzerland)*

"Too often our focus as management educators is only on the content—the models, concepts, tools, and techniques – to solve challenging problems. What if we could go deeper? What if we could provide experiences that enable learners to fundamentally shift their beliefs about society, organizations, and people, especially themselves, and to convert these beliefs into meaningful action? The contributions in this volume have been carefully curated to help us do exactly that. It is for educators who want to go deeper, do better, and make a bigger difference in building a better world."

**Dan LeClair**, *CEO, Global Business School Network (USA)*

"Finally! A book that gives practical tools for people who want to change the hearts and minds of students in sustainability. This book offers insights from sustainability instructors from around the world and will enrich any curriculum. Learn about simulations, experiential learning, and case studies that will transform individuals and stimulate action."

**Tima Bansal**, *Professor, Ivey Business School; Executive Director, Network for Business Sustainability; PRME Board Member (Canada)*

"The book is a great source of inspiration and a great collection of insights into the potential of business education in preparing a truly new generation of managers for a sustainable future. It is a clear demonstration that we are witnessing a shift in values that is changing management education and management research and will ultimately change the role of business schools in improving society."

**Sergey Myasoedov**, *Vice-rector RANEPA University, Director/Dean IBS Moscow of RANEPA, President of RABE (Russia)*

"This is a wonderful book. United in their quest to make a better world, faculty on five continents gather here to share their best teaching ideas. Even better, we hear from their students. It is great to discover just how much they care. The world will be a better place if we follow their lead."

**James P. Walsh**, *Carey Professor of Business Administration, Ross School of Business, University of Michigan; Former President, Academy of Management; PRME Board Member (USA)*

"A celebration of perspectives! *Revolutionizing Sustainability Education* offers mind- and heart-opening insights into the journeys and practices of students and educators around the globe. A true inspiration for what is possible in the now – and what we can envision for our collective future."

**Sophie Charrois**, *President, oikos International (Switzerland)*

"A refreshing take on the *management* curriculum. For too long, we have educated our future 'leaders' on how they can best manipulate circumstances in order to achieve some external goal. In doing so, we have taken their attention off of the only thing that matters, which is the Self. Try to control everything, and you will end up controlling nothing. Focus on understanding the Self, on the other hand, and you stand a chance at mastering your entire existence. This book is an encouraging step in that direction."

**Duff McDonald**, *author of* Tickled: A Commonsense Guide to the Present Moment *(Harper, October 2021)* *and* The Golden Passport: Harvard Business School, the Limits of Capitalism, and the Moral Failure of the MBA Elite *(Harper Business, 2017) (USA)*

"This book provides a wealth of insights into the craft (or art?) of sustainability education. It demonstrates that traditional approaches to teaching sustainability, which aim at imparting knowledge in students, have clear limitations; to facilitate deep-level learning and provide truly transformational experiences, a more holistic approach to teaching sustainability is needed that promotes learning at the intellectual, behavioral and affective levels."

**Günter Stahl**, *Professor of International Management, Director of the Centre for Sustainability Transformation and Responsibility (StaR), the Vienna University of Economics and Business (WU Vienna) (Austria)*

"If you want learners – at whatever level – to engage more deeply with sustainability issues, this book provides the tools and approaches to do just that. Using classroom-tested approaches that really transform thinking about sustainability. Rimanoczy and Ivanova capture how innovative instructors facilitate mindset shift through active learning."

**Sandra Waddock**, *Professor of Management, Boston College; Author of* Transforming towards Life-Centered Economics *(Business Expert Press, 2021) (USA)*

"Dr. Ekaterina Ivanova and her colleagues have written an intriguing and interesting book that emphasizes the crucial link between economic efficient activity and its aesthetic and eco-ethical ramifications. Definitely not mainstream management practice but likely a necessary component to give back some elements of purpose to business. A book that should be read for anyone who believes that business needs to 're-source' itself, to embrace and embed itself in a broader and more holistic and eco-aesthetic world."

**Peter Verhezen**, *Visiting Professor, University of Antwerp (Belgium)*

"What a wealth of unique learning methods one can find in this book! I would recommend it as an inspiration to anyone who wants to change their and/or others' mindsets to be(come) a role model for the type of human being our species needs to survive on this planet. Beyond inspiration the book provides hands-on guidance and the experiential accounts of the ones who 'have done it': It can be done, because it has been done."

**Oliver Laasch**, *Ph.D., Senior Lecturer in Entrepreneurship and Innovation, Alliance Manchester Business School, The University of Manchester; Author of* Principles of Management: Practicing Ethics, Responsibility, Sustainability *(Sage, 2021) (UK)*

"This inspiring book offers insights, ideas, and examples of how education can contribute to profound shifts in the way we think about and relate to global sustainability. It highlights the power of shifting from teaching to facilitating, and shows us that transformations are possible – and occurring."

**Karen O'Brien**, *Professor, Department of Sociology and Human Geography, University of Oslo (Norway)*

"There is a prevailing – and incorrect – view within higher education that we must merely provide students with facts and knowledge and they will be equipped to solve the great challenges of the day. That is not working. What is missing is a call to be inspired with a sense of hope, challenged to discern their personal purpose and assisted in striving for the higher aims of a career worth living, one that contributes to something much bigger than themselves, the common good. This book is an invitation to rethink how we educate, provoking a reexamination of our roles as professors and students. We must stop simply teaching and start guiding, facilitating, mentoring, inspiring and at times co-learning with our students to transform our mindsets to tackle problems for which the old mindsets are now obsolete. If you care about teaching and about the fate of our world, read this book."

**Andrew J. Hoffman**, *Professor of Sustainable Enterprise, Ross School of Business, University of Michigan, Author of* The Engaged Scholar: Expanding the Impact of Academic Research in Today's World *(Stanford University Press, 2021) and* Management as a Calling: Learning Business, Serving Society *(Stanford University Press, 2021) (USA)*

"My journey with the sustainability mindset started by joining the PRME Working Group on SM, and joining the SM Fast track (online) workshop with Isabel Rimanoczy. It was an eye opener to see an approach that treats sustainability from a combined holistic and humanistic view. I had the privilege to interact with Isabel again and with Ekaterina in the Atelier about SMP; a virtual meeting place with a limited group of participants to discuss the SMP and to bring those principles to students. This book is a capstone to many of the ideas discussed in the Atelier and more activities of the two authors and many more. I can appreciate how the concepts in the book were developed through interaction and validation with a large community immersed into the topic of sustainability through teaching and research. I encourage you to read this book and to start implementing the concepts in your classroom. We need to start a movement to instill the sustainability mindset in all students of sustainability, and to encourage them to be the change they wish to see in the world (paraphrasing Gandhi's quote)."

**Ali Awni**, *Professor of Practice, Director of the John D. Gerhart Center for Philanthropy, Civic Engagement and Responsible Business, The American University in Cairo (Egypt)*

"This timely and innovative book not only acknowledges the importance of having a whole person approach towards management education, but also reflects the power of 'slow learning' and experimentation. Such perspectives are urgently needed to engage students in the challenges of tackling the world's most pressing sustainability challenges and enable them to realize their potential. Through this book, Rimanoczy and Ivanova provide an intriguing testimony of how management students and educators alike can give voice to their feelings, values, and ideas in a time of great transition. I think it is a highly commendable read."

**Lars Moratis**, *Deputy Professor in Corporate Social Responsibility, Antwerp Management School (Belgium)*

# REVOLUTIONIZING SUSTAINABILITY EDUCATION

There is growing awareness among leading responsible management scholars and practitioners that understanding global wicked problems is insufficient in effecting lasting engagement and changed behaviors. Research indicates that to impact behavior, the mindset has to shift, which leaves the question: How do you shift a mindset?

This book guides educators and practitioners, their students and colleagues to take action on finding urgent solutions to the grand challenges stated in the 17 UN Sustainable Development Goals. A Sustainability Mindset is a way of thinking and being that results from a broad understanding of the ecosystem, from social sensitivity and an introspective focus on our personal values and higher self, which finds its expression in actions for the greater good. By promoting a mindset shift, educators in very diverse contexts are laying the foundation for a resilient future. The book presents a collection of over 150 student voices depicting a transformative experience and a shift in their mindset. Seventeen educator/student teams of contributing authors from across five continents describe the activity that prompted those students' reflections, and the conceptual frameworks that played a role in the selection of the learning goals and activities.

The book is written with academic and corporate educators, reflective practitioners, consultants, coaches, trainers and students in mind, and is invaluable in guiding the process of developing a sustainability mindset among participants in the training process.

**Ekaterina Ivanova** is an Associate Professor in the Department of Strategic and International Management at the Graduate School of Business at the HSE University in Moscow. She has over 15 years' experience as an applied and academic researcher, non-profit executive, and educator focusing on ethics, aesthetics, responsible business and sustainability mindset. As a co-chair of the PRME Working Group on the Sustainability Mindset, AIM2Flourish professor (Flourish Prizes 2021 for the SDGs 4 & 17) and founder of the Sustainability Navigator Telegram Channel she is creating an impact for a better world through knowledge transfer, outreach and action.

**Isabel Rimanoczy** is author of *Big Bang Being* (2014), *Stop Teaching* (2016) and *The Sustainability Mindset Principles* (2021). She is the Convener of the PRME Working Group on the Sustainability Mindset, a cohort of over 170 academics from 165 universities on five continents promoting innovation and a sustainability mindset. Global Ambassador of AIM2Flourish and Fellow of Schumacher College, more recently, she participated in creating the Sustainability Mindset Indicator, a personal development tool for individuals and educators.

## The Principles for Responsible Management Education Series

Since the inception of the UN-supported Principles for Responsible Management Education (PRME) in 2007, there has been increased debate over how to adapt management education to best meet the demands of the 21st-century business environment. While consensus has been reached by the majority of globally focused management education institutions that sustainability must be incorporated into management education curricula, the relevant question is no longer why management education should change, but how.

Volumes within the Routledge/PRME book series aim to cultivate and inspire actively engaged participants by offering practical examples and case studies to support the implementation of the Six Principles of Responsible Management Education. Books in the series aim to enable participants to transition from a global learning community to an action community.

Books in the series:

**Inspirational Guide for the Implementation of PRME: Placing Sustainability at the Heart of Management Education**
*Edited by the Principles for Responsible Management Education*

**Global Champions of Sustainable Development**
*Edited by Patricia M. Flynn, Milenko Gudic and Tay Keong Tan*

**Unmasking Irresponsible Leadership: Curriculum Development in 21st Century Management Education**
*By Lola-Peach Martins and Maria Lazzarin*

**Struggles and Successes in the Pursuit of Sustainable Development**
*Edited by Tay Keong Tan, Milenko Gúdic, and Patricia M. Flynn*

**The Sustainability Mindset Principles: A Guide to Develop a Mindset for a Better World**
*Isabel Rimanoczy*

**Business Transformation for a Sustainable Future**
*Edited by Samuel Petros Sebhatu, Bo Enquist and Bo Edvardsson*

**Revolutionizing Sustainability Education: Stories and Tools of Mindset Transformation**
*Edited by Ekaterina Ivanova and Isabel Rimanoczy*

**Responsible Management Education: The PRME Global Movement**
*Edited by Mette Morsing*

# REVOLUTIONIZING SUSTAINABILITY EDUCATION

## Stories and Tools of Mindset Transformation

*Edited by Ekaterina Ivanova and Isabel Rimanoczy*

Routledge
Taylor & Francis Group
LONDON AND NEW YORK

First published 2022
by Routledge
2 Park Square, Milton Park, Abingdon, Oxon OX14 4RN

and by Routledge
605 Third Avenue, New York, NY 10158

*Routledge is an imprint of the Taylor & Francis Group, an informa business*



*British Library Cataloguing-in-Publication Data*
A catalogue record for this book is available from the British Library

*Library of Congress Cataloging-in-Publication Data*
A catalog record has been requested for this book

ISBN: 978-1-032-13537-3 (hbk)
ISBN: 978-1-032-13538-0 (pbk)
ISBN: 978-1-003-22973-5 (ebk)

DOI: 10.4324/9781003229735

Typeset in Bembo
by Deanta Global Publishing Services, Chennai, India

*To every single teacher in my life, but especially to my father Alexander and my husband Valerian for their eternal love, faith and inspiration in convincing me to take the path of an educator.*

*–EI*

*To the many pioneers who had the courage to question what is, and the imagination to envision what could be. We are standing on your shoulders.*

*–IR*

# CONTENTS

# ILLUSTRATIONS

## Figures

## Tables

# CONTRIBUTORS

**Royina Bakshi (Australia)** is a Masters of Management (Marketing) graduate from the University of Melbourne. She currently works as a Public Relations and Marketing Specialist at Melbourne-based medtech startup Nutromics. In the past, she worked with Goldman Sachs, India as a Senior Risk Analyst.

**Armand Bam (South Africa)** is Head of Social Impact at the University of Stellenbosch Business School in South Africa and Senior Lecturer in Business in Society on the MBA program. He is responsible for the Social Engagement Project (SEP), an experiential learning assignment, aimed at providing a transformative learning experience to MBA students.

**Elina Hanhimäki (Finland)** is a biology Master's student at the University of Oulu (Finland). As a science student she thinks that scientific research gives rise to new sustainable solutions and business opportunities.

**Brian Hanssen (USA)** is the Associate Director and Clinical Assistant Professor of Management Communication at the NYU Stern School of Business. Previously, Brian was a consultant specializing in the effective communication of large-scale organizational change/ social impact initiatives. He holds degrees from UCLA and Columbia University.

**Veera Hartikainen (Finland)** is a Master's student of molecular medicine in the Faculty of Biochemistry and Molecular Medicine at the University of Oulu (Finland). She wants to create more sustainable innovations and job opportunities.

**Leon Hecht (Germany)** studies business administration and engineering at the RWTH Aachen University in North Rhine-Westphalia (Germany). His interest in sustainable entrepreneurship originates in the complexity of measures taken by economical actors and entrepreneurs in order to adapt to a market that, due to rising environmental awareness and urgency of the issue, is evolving rapidly.

**Jan Hermes (Finland)** is an Assistant Professor at Oulu Business School (Finland) and a university professor at Nicolaus Copernicus University in Toruń (Poland). His research interests are in organizational institutionalism and business responsibility. Jan's research has been published, among others, in International Journal of Management Education and Social Responsibility Journal.

**Iris Maria Hertog (Sweden)** works as a junior process manager at NextGreen, a small consultancy in the Netherlands that facilitates multi-stakeholder processes on sustainability. Iris holds a Bachelor in Forest and Nature Conservation (Wageningen University, the Netherlands) and a Master in Environmental Studies and Sustainability Science (Lund University, Sweden).

**Ayako Huang (USA)** is an Associate Professor and the director of the MBA in Sustainable Business at Maharishi International University, Fairfield, Iowa. Her work focuses on education, learning and development of competencies for sustainability in both academic and organizational environments.

**Martin Friesen (Germany)** studies business administration/ controlling finance and accounting at the University of Applied Sciences in Pforzheim (Germany). His motivation to get engaged in sustainable entrepreneurship comes from his curiosity about how to set up and develop a healthy and sustainable business.

**Morgane M.C. Fritz (France)** is Associate Professor in Sustainability Management in Supply Chains at Excelia Business School, La Rochelle, France. Her research and teaching interests include supply chain sustainability, sustainable procurement, business ethics and stakeholder management. She is published in and acts as a reviewer and guest editor for various peer-reviewed journals.

**Amelia Naim Indrajaya (Indonesia)** is teaching at IPMI International Business School, Jakarta, Indonesia. She is the founder of Center for Sustainability Mindset and Social Responsibility. She obtained her doctoral degree from Strategic Management, Universitas Indonesia and got her MBA degree from the University of Colorado at Boulder.

**Michael Lees (USA)** is a Professor of religion, philosophy and creativity. He received his PhD in Global and Comparative Education. He offers 25 years of college teaching experience. His research interests include examining the intersections of contemplative and spiritual ecoliteracy approaches to learning for emerging adults and educators in higher education.

**Sini Liimatainen (Finland)** studies economics and business administration at Oulu Business School (Finland). Her interest in sustainable business comes from the desire to combine her studies and personal life, where she strives to live as sustainably as possible.

**Lasse Lychnell (Sweden)** (Ph.D.) is an affiliated researcher and a teacher at the Stockholm School of Economics (SSE) and is part of the team that created the mandatory Global Challenges track launched in 2016. He has primarily focused on developing a sustainability mindset by integrating theory, practice and subjective experience through contemplative exercises.

**Hanna Małaszkiewicz (Poland)** studies management at the Warsaw School of Economics (Poland). In sustainable entrepreneurship, she sees the opportunity to create and spread real values that are able to ensure the multidimensional development of society and bring us a better future.

**Emma Le Maître (France)** studies business and marketing at Kedge Business School in Bordeaux (France). She considers sustainable entrepreneurship the main challenge of tomorrow as our habits and society change and need to be sustainably understood and reviewed. So, it is up to us.

**Catarina Neves (Portugal)** studies chemical engineering at the University of Porto in Porto (Portugal). Her interest in sustainable entrepreneurship was born from the desire to set up her own business coupled with her admiration for the natural symbiosis of the ecosystems.

**Manthan Pakhawala (USA)** graduated in 2021 from the NYU Stern School of Business (Psychology minor). He currently works with McKinsey and Company as a Solution Delivery Analyst. You can find Manthan obsessing about the latest Christopher Nolan film, rooting for FC Barcelona, or re-watching The Office for the millionth time.

**Lucia Di Paola (Sweden)** lives in Germany where she works as a Just Transition officer at ICLEI Europe, a global network of local governments committed to sustainable urban development. She is passionate about environmental justice and in the role of inner transformation in the just ecological transition.

**Rocky Reynaldo (Australia)** is a Masters of International Business graduate from the University of Melbourne and a recipient of the prestigious Australia Award ASEAN Scholarship. He currently works as an Associate Recruitment Consultant at Michael Page in Indonesia. Previously, he worked in the retail and music education sector.

**Isabel Rodríguez-Tejedo (Spain)** is a Professor of Economics at the School of Economics and Business of the University of Navarra. She obtained her Ph.D. from the University of Maryland and has been an invited researcher at the GW Institute of Public Policy, the Inter-American Development Bank and the World Bank.

**Sheila Sampaio (Brazil)** is a Professor at Sonata Brasil. She holds a Master's in Management and Business from Université de Poitiers – France. Professor of Leadership and Human Connections in postgraduate courses. She received the Top Human Being Award 2019 (ABRH – Brazil) for her master's dissertation on Leadership and Spiritual Intelligence.

**Abigail B. Schneider (USA)** is an Associate Professor at the Anderson College of Business and Computing at Regis University in Denver, Colorado. She earned her Ph.D. in Business Administration (Marketing) from the University of Colorado at Boulder and her B.A. from Colgate University. Her research focuses on consumer/societal/environmental well-being.

**Arnold Smit (South Africa)** is Associate Professor of Business in Society at the University of Stellenbosch Business School in South Africa. The focus of his teaching, research and social engagement is on the integration of ethics, responsibility and sustainability in management education, leadership development and organizational practice.

**Karthyeni Sridaran (Australia)** works as a sessional academic in Management at the Faculty of Business, University of Melbourne, Australia. She has over 25 years of international work experience in public, private, consulting and academic sectors. Her areas of interest are in conscious leadership, workplace wellness, positive psychology and sustainable mindsets.

**Karol Vieker (Sweden)** has over 15 years' experience of diversity, equity and inclusion issues related to working life and the educational sector. She is passionate about DEI and has dedicated her working life to making the world a better place by striving to create open and inclusive environments.

**Julia Storberg-Walker (USA)** is Department Chair and Associate Professor in the Department of Human and Organizational Learning, George Washington University. She received her Ph.D. in Work, Community, and Family Education from the University of Minnesota. Her current research and practice interest lies at the nexus of leadership, consciousness, equity and planetary flourishing.

**Christine Wamsler (Sweden)** is Professor of Sustainability Science at Lund University Centre for Sustainability Studies (LUCSUS) and director of the Contemplative Sustainable Futures Program. She is an internationally renowned expert in sustainable development and associated (inner and outer) transformation processes with more than 20 years of experience, both in theory and practice.

**Deyang (Mike) Yu (USA)** is a doctoral student at George Washington University who is passionate about social sustainability. He completed his bachelor's degree in Economics at University of Washington, and he earned his master's degree in Finance from Johns Hopkins University. After serving in the banking industry for years, he returned to explore the essence of sustainability education.

# FOREWORD

As I enter the later stages of my career, all of which I have spent in higher education, I often wonder about the impact of what we university professors actually do. I worry that the university is set in its ways as a medieval institution that is very difficult to change, and slow to react to the pace of change in the modern world.

This wonderful book reaffirms my natural optimism that we can and often do make a real difference with our students. The 17 essays/ stories in this volume from all over the world recount many ways that a group of professors have affected the lives of their students for the better. And, I want to emphasize that they have affected their lives, not just their conceptual ideas.

The authors have created a platform for experiential learning around sustainability, but the platform could have been around many other topics, and I suspect, or hope, that there are other groups of academics in other disciplines that have done a similar task. The book is built on intense student engagement, an idea that is often found wanting in university lecture courses that are still, unfortunately, the norm.

Experiential learning becomes transformational when the students can see and reflect upon how these experiences can affect their lives. The student reflections in the essays are worth the price of the book themselves. We can listen to what these students say, and we can see how just a set of good lectures pales by comparison to real experiences. One professor has developed a way to connect emotional intelligence and sustainability, thereby echoing John Dewey's idea that we have to "get inside a problem" especially if it is an ethics problem like sustainability. Another professor from the US takes students to Uganda so that they can have some experience about a very different society. And, these are just two of a plethora of ideas.

The creators of these stories are not mired in the old story of business that business is just about profits and the money. They are committed to the idea that good business can change the world, that business has to create value for its stakeholders, and that sustainability is an urgent issue that speaks to all businesses and business people.

In summary, this book will repay reading and re-reading the essays, in order to join the party of those who are creating cutting-edge methods and courses to transform learning experiences. My sincere thanks to all involved.

*R. Edward Freeman*
*University and Olsson Professor*
*The Darden School, University of Virginia*
*Charlottesville, Virginia, USA*
*31 May 2021*

# INTRODUCTION

This book you are reading started with a small but meaningful event. It was during a regular semester pre-Covid, when I (Isabel) was reading the reflective assignments of my students at the Universidad de Navarra, Spain. I was surprised to notice that several students had written their assignments in poetry (without being asked to), or at least in very poetic prose. Since the students were part of a management course focused on understanding self and the contemporary world and not a literary class, I paused in awe at their very profound (and unusual) ways of expressing their thinking.

My course had been brief, yet the way the students described their transformational learning experiences was astounding, both in form and in content. I was moved to share these exceptional pieces with my colleagues of LEAP, the Global Compact's PRME Working Group on the Sustainability Mindset, as an inspiration of the transformational learning we can create. I discussed it with Ekaterina, the idea came up: what if we created a book with students' voices, where professors could select a few paragraphs or fragments depicting the impact their teaching had on their students, as they were developing a Sustainability Mindset? As we were listening to the presenters of the plenary session of the annual meeting of Academy of Management in Chicago right after our annual retreat, we felt that this was the right moment to start working on the exciting book project.

## How are learning facilitators transforming minds and souls

I (Ekaterina) immediately fell in love with the concept! I had been experiencing myself how students in my courses on business ethics and responsible business at the Higher School of Economics in Moscow were often puzzled with my non-conventional pedagogical approach. It is based on loosely defined creative tasks to address the global grand challenges with a focus on experiential learning, contemplative practices and practicing shared leadership in groups. Despite being taken out of their comfort zone, students were able to reach powerful insights and explore new worldviews. So we quickly put together an invitation to our colleagues in the Working Group, asking if any others had prompted some transformational learning experiences with their students, and were willing to describe the activity or assignment that had played a major role in the students' mindset shift. We also were interested in the students' words, and determined to include key portions of their essays describing their experience. We wanted the book to be a co-creation that shows the process of learning as a transformational journey for both the students and their learning facilitators.

Big was our surprise as we received 17 responses in less than one week. Colleagues from Latin America, Africa, Asia, Europe, North America and Oceania expressed their enthusiastic interest in telling their stories. The more we read their stories, the more convinced we were that something significant was taking place around the world, and that it had to be documented and made public. Most of our colleagues teach in more or less traditional institutions, and lead typical courses such as Business Ethics, Responsible Business, Business in Society, Global Challenges, Global Political Economy, Management and Organization, Management Competencies, Sustainability and Inner Transformation, Entrepreneuring for Sustainability, Leadership, Marketing for Social Change, Social Impact Scholars and Supply Chain Management. What they have done differently, though, is that they have adapted their syllabi to teach in a more holistic way, one that engages the whole person, not just the mind, of their students. They are trying out different pedagogical approaches, acting at times more like learning facilitators and just-in-time teachers, rather than mere instructors. They are bold and courageous change-makers transforming their courses through active engagement with the students, taking them out of their own comfort zone, and connecting their heads, hearts and hands. They are bringing into their classroom novel activities that make their students pause, develop self-awareness, reflect and express their thoughts through art or improvisation, contemplative or social practices, actual or imaginary traveling experiences that lead to a deeper connection with the universe. They challenge self-exploration among the students by creating safe environments where students can feel engaged, inspired to take action, give weight to their dreams, identify their talents, revise their values and paradigms – all the anchors of their identity. They use more questions than answers, and in many cases create a slower pace in the room, conducive to more profound insights. They find fun and innovative ways to connect the discussions with the world, with the social and environmental challenges, and with what each of us can do.

We came to believe that the book was not just a great idea, but that it was also our responsibility to create a platform to showcase these powerful events. The stories address the cornerstone of our pedagogical challenge, well described by Stephen Sterling, Guy Dancey and many other scholars in the discussion theme "Educating for the future we want."[1] Education is far overdue for a change to meet our current reality. We need to revise what we teach, what we want to achieve, what skills, competencies and mindset the students will need in order to be valuable shapers of a better world.

And how we will do that? To us, the unassuming and simple stories from Sweden, Finland, France, Spain, Russia, Indonesia, the USA, Brazil, South Africa and Australia in this book precisely attest to these new pedagogical approaches, the holistic engagement of the individual, inquiry-based ownership of the learning by the student, the transformational experiences and the development of a new mindset shift that can result from the changes.

## How this book is organized

We grouped the stories that collectives of 17 teams co-created into two categories – individual transformational experience, and action on the SDGs. The first part focuses on individual transformational experiences, featuring a collection of students' reflections depicting a transformative experience and a shift in their mindset. Their instructor, in turn, describes in each case the syllabus or type of activity that prompted those students' reflections, and any conceptual frameworks that played a role in the selection of the learning goals and activities. With essays and exercises from around the world, the authors show that the shift towards a sustainability mindset is possible in very diverse contexts.

The second part focuses on the mindset shift driving action on the Sustainable Development Goals. It is clear that the role of educators has to be increasingly aimed at facilitating learning that results in action. The contributing authors understand that a mindset becomes evident in actions taken for the greater good. As such, this part provides educators and leadership development professionals' inspiration of the change they can effect, through the testimonies of participants, underpinned by the description of the learning tools and concrete activities that made it happen. The contributors are convinced that the time for acting is now because the future is being created in the present moment. And they show how.

## How the book can be used

Each story presents a particular assignment or activity, describes the context in which it took place, provides sufficient detail for any reader to replicate it, and cites the students' powerful excerpts as a testimony of the impact. The contributing authors also welcome questions and provide their contact information so as to continue the dialogue.

The collected learning tools have been used at the universities and business schools that are part of the PRME global community, or are considering joining it. The learning tools have been applied across the undergraduate and graduate programs, in the degree and non-degree formats, in offline, online or blended formats, and in mandatory or elective settings. Contributions include insights about the students' engagement, challenges encountered, surprises and references to other activities that have contributed to the mindset shift. A number of contributions tell follow-up stories describing what the students did afterwards in connection with, and as a result, of a shifted mindset.

We have one major objective for this book: that it serves as a practical guidance for academic and corporate educators, consultants, coaches and reflective practitioners around the world teaching courses and facilitating workshops on responsible, resilient and sustainable business practices. We want it to be a source of inspiration and a recipe book ready for immediate use in leadership programs and classrooms. Given the diversity of contributors from around the globe, we hope it will enable learning and education practitioners to find assignments and activities suitable for their own context.

You will find in these pages a fractal of the world. You will hear a polyphony of voices from 17 learning facilitators and over 150 students as they discuss their passion, concerns, emotions, feelings, dreams, confusion, questions, trial-and-learning, authenticity and self-discovery. You will find unrehearsed words coming from the soul. We kept the exact written comments used by the students in not always perfect English. Most of the students quoted in this book speak several languages, with English being just one of them, and frequently the second or third they learned.

We also hope that you will find inspiration and that you will be excited to try out some of these activities, or to contact the contributing authors to find out more. If you are interested in learning more about the theory and conceptual framework of the Sustainability Mindset, you will find a list of resources at the end of this book. We called this book "Revolutionizing Sustainability Education: Stories and Tools of Mindset Transformation" because we believe that you can benefit from the practical experience collected in this book and find your own path for a better world.

Enjoy the discovery!

*Isabel Rimanoczy and Ekaterina Ivanova*

## Note

1 https://greattransition.org/gti-forum/education-sustainability-sterling.

# ACKNOWLEDGMENTS

Continuing the tradition of LEAP! members embarking on a new book project, this one turned out to be yet another "love child." It helped many of us stay connected and, together, overcome the challenges of the COVID-19 pandemic. Our first thank you goes to the supportive community of 170+ academics who are members of the LEAP! Family – the PRME Working Group on the Sustainability Mindset – for simply being there, for listening, and for engaging in the work we are doing on behalf of revolutionizing sustainability education around the world. Since its inception in 2013, LEAP! (acronym for Leverage resources, Expand awareness, Accelerate change and Partner) and its community grew organically, connecting nonmainstream educators from all the continents. LEAP! members are inspired by the mission of transforming teaching practices to help students become the best versions of themselves. We are creative pioneers inviting our students to connect their minds with their hearts and hands. This book, with 17 contributing teams of educators and students, is representative of the impact of LEAP! on the lives of those we reach.

We sincerely appreciate the support of the PRME Secretariat that we received from Mette Morsing, Nikolay Ivanov, Sophie Kacki, Luisa Murphy and our peers from the PRME Working Groups, especially Milenko Gudic for his support, mentoring and encouragement.

Our deepest appreciation goes to Rebecca Marsh for her trusting in and championing our book from ideation to publication. We also would like to thank the many anonymous peer reviewers for recognizing, in this book, the value of the stories and tools shared by students and educators around the globe.

Since most of our contributors are non-native speakers of English, we are grateful to Martine Marie, Veronica Legrand and Chris Murray for their efforts to polish our English. We are also pleased to acknowledge the work of Claudine Brunnquell who helped us finalize the manuscript.

Our highest gratitude is reserved for each contributor, for their professional work, dedication, patience and belief in this, our book project.

*Ekaterina Ivanova and Isabel Rimanoczy*

# PART I

# Individual transformational experience

# 1

# JUST BE THERE

## The inconvenient exercise of being with oneself and doing nothing

*Isabel Rimanoczy*

### Prologue

We design courses with the outcomes in mind. So I want to start this chapter with the outcome reached, in the voices of the students. These three essays were written by undergraduate students[1] for an assignment given in a face-to-face course called "Understanding Self and the Contemporary World," that I taught in Spain and in Morocco in 2019. I decided to start this chapter with their words, because when I ask myself why I am an educator, my personal answer is to provoke, foster, support and accelerate the personal transformation of the mindset.

*Student reflection 1, female, Spain*

*I waited the whole week for a day with good weather to do this adventure, but as days went by and the sunny days didn't come, I decided to face the exercise with bad weather.*

*I don't have a car, so I couldn't go to a fully natural place. I walked up to a place close to my home that could be seen as nature, a pine forest a bit away of everything.*

*I thought it was a nice exercise. There were some people walking their dogs, and I thought it would be nice to have a dog, that would oblige me to walk in nature at least a few times a day. Also, with a dog I would feel accompanied, I don't like to be alone.*

*I was anxious because I didn't feel the weight of my mobile on me. I felt unprotected, naked. At the same time, I felt very light, as if I wouldn't be carrying anything. In fact, I was not carrying anything, but I don't know how to explain it.*

*The silence felt awkward. It was disquieting in the beginning, I didn't know what to do. Then I started to get used to. I realized that I never am in silence, I am always listening music, or class, or talking to someone, or studying (although studying happens in silence, it's not a real silence because your brain is going at 100%).*

DOI: 10.4324/9781003229735-2

*I noticed that I almost, almost was afraid to be alone with myself. What should I do next? How boring. I thought of what everyone says these days, we hear a lot that we have to learn to be alone with oneself. This is something I haven't mastered, but I would like to, I think it's important. Perhaps exercises like this one help me. But I also think it's difficult and almost impossible to go on your own account to a place in nature, without anything and anyone. I have the feeling that it is an exercise you wouldn't do until the day you are very unhappy, with yourself and everything, and you seek a way to escape, and so you finally do what everyone has told you: mindfulness, meditation, and time in nature … However, it's a paradox. Because probably (I actually don't know for sure), if meditation, nature, to be with yourself, appreciate silence, the fresh air … would be part of your routine, you might never get to the point of feeling bad and seeking an escape.*

*I enjoyed disconnecting from everything for a little while, but I also realized how dependent I am with my mobile, and I thought it's horrible, I would like to change that. I thought it would be also good to go out without my watch, I also depend so much on my watch. But that really would have made me feel naked and super lost. Step by step.*

…

*Well, to end this I would say that this activity has brought me peace. More than anything. What we all need these days is to have a time to meditate, to reflect, to be quiet. If that time is in nature, it's a plus. However I believe that if we could do these small moments of meditation and step away from our mobile phones and everything, for a little while, be it in nature or where you can, we all would get this experience of peace. And with this feeling of peace many other things come along, like reflecting on what is important and what is not, a moment to stop and process what we are doing, and what is happening to you, peace with other people … And I think that truly, if someone is happy and at peace with oneself, it's easier to have a sustainable way of life.*

*Student reflection 2, female, Morocco*

*In order to fulfill the requirements of this experiment, I have undergone 1 hour of meditation in a garden that is 10 minutes from where I live, deprived of social interaction, technological distraction or any form of recreational activity. This experience was quite unique for me and was therefore a bit enlightening. In the lines to follow I will elaborate more on my thoughts and feelings towards this, dare I say, adventure.*

*The prologue of this experiment was a bit rough, due to the unusual nature of the task at hand. I'd like to address some of problems I ran into, what I learned from them and what my overall thoughts on this experiment are.*

*I will start with the very first issue that I encountered: the disconnection. While I did enjoy the peaceful, tranquil environment that surrounded me; the cool summer breeze was especially welcome, I did feel a sense of "anxiety" overcome me. I found myself creating different tragedies and highly disturbing scenarios that could all transpire in the absence of my phone. While I must admit that I entertained those tall tales for longer than I initially anticipated, I was ultimately able to dispatch them from my consciousness. This allowed me to focus more on the moment I was experiencing, which leads me to my next point: the lack of stimuli.*

*Saying my first impression of this meditation was "dull" would be an understatement. It was difficult to enjoy just being there. This was quick to change, after taking a deep breath, listening to my surroundings and emptying my mind from the previous negative premonitions, I suddenly discovered it was rather relaxing.*

*The serenity of nature along with the rustling of the trees was therapeutic, a stark contrast to the hustling lights and noises I was used to in my daily life. I believe that my view of such an activity was tainted by the fashion in which modern lifestyles have evolved. At every waking moment we find ourselves bombarded by visual cues or auditory ones leading us though our days. The constant beeping and buzzing of our phones urging us to respond to a conversation we may not want to be a part of us, signals for emails that pry us out of our breaks and remind us pending deadlines and the overlapping sense of urgency in our every move, loud music that drowns out our very own thoughts making it impossible to truly think and reflect on our actions etc. all those factors, in my opinion, relate to one simple idea: we're more connected to the outside world, but not to ourselves. In that brief hour I spent away from those distractions, despite the initial setbacks, I felt truly at peace. I think that we've learned to busy ourselves so much that we've forgotten how important it is to, sometimes, have nothing trouble us. In that 1 hour there was no rush no real obligation and I think that would be something we should all experience regularly in order to be healthy both mentally and physically.*

*In conclusion, I'll say that this experiment was eye opening for me and that it helped me shed some light on myself and the environment I live in. The overwhelmingly busy lifestyle of a student is always something that people try to deal with, but I think that this experience has helped cope with some aspects of it. It made me seriously consider making this a regular habit and I believe everyone should at least give it a try.*

*Student reflection 3, female, Spain*

*I just had a totally new experience. I say new because I cannot recall when it was that I last was able to disconnect from the routine and to re-connect with myself and with the world. It will be very difficult to put into words all the reflections and feelings that crossed my mind during these 45 minutes. Now that I look at my watch, it probably has been more than that, I think I changed the rules a bit of this game. I will start with what I feel was the main trigger for my reflections.*

*I remember stepping out the door thinking of this exercise as an obligation. All kinds of negative thoughts crossed my mind, towards the people that had taken the liberty to rob me of 45 minutes of my time, in a period so complicated like this one. With the finals around the corner, the grades I have to maintain to keep the conditions that allow me to study at this school, and the pressure to finish the course so that I can meet the expectations of the company that just hired me … I don't have time for this silly thing, and now less than ever!*

*I have to smile as I recall my string of thoughts. I came back through this door convinced that someone has given me the biggest gift I have received in a long time. Over the past five years I have thought that people outside of my life were imposing on me a routine of obligations that it was impossible to escape from. Today, I realized that it was me who was converting the routine of not stopping into an addiction. Addictions are bad. Like addiction to tobacco kills you, the addiction to not stopping kills you too. For this reason, today I say I have received a gift.*

*After getting my mind to calm down and let go the rebel attitude, I just observed. We have been talking about sustainability and the planet for several days, how it affects us and what we can do.*

> *We looked at the human being as an external agent. We empathized with vegetation, animals, soil, etc and instinctively we had negative feelings towards human being. The human being is guilty and is the solution, at the same time. The first part is fine, but the human beings of the first world is also part of it. I am thinking of a new concept: the sustainability of people.*
>
> *No one knows what I am doing here, sitting. Strangers walk by, some seem to be enjoying the sunny day going for a walk; others seem to be in a rush, perhaps they arrive late to the bus stop; there are a few sitting on the lawn, others are playing with a Frisbee. But I can only observe how they are physical, and imagine their stories. Each one has their story, and I have mine. And on top, I am too worried with my own issues, which probably are more complicated than anyone else's. What a lie. A big fat lie.*
>
> *The addiction to the routine of not stopping makes us selfish. It threatens our natural condition as social beings. We are so worried about our issues, that we forget there are others around us, we become individualistic, the main ingredient of sadness. Sustainability of people is essential for the sustainability of the planet. This sustainability starts with stopping to be pre-occupied and start occupying.*
>
> *I want to yell into the wind, starting with myself, so that we all observe nature without the need to go for a walk. I mean to really see, to listen, to notice. Without the need to go for a walk means that nature is in the people with whom we share our obligations every day. We have around us stories that need our attention. I think the first grain of sand at our reach for the sustainability of the planet starts with the people around us. If we all collaborate in this, we will recover our social nature, and in this is embedded the sustainability of the planet. Because the care for the planet is embedded in us, but we have stopped being "us" and have become an addition of 7350 million of egos. The planet's sustainability starts with no one feeling alone, because each one of those 7350 million persons is because it's a part of the others.*
>
> *I would like to finish expressing my gratitude for this gift I received. It's a pity that the word "thank you" is a bit worn out, because of excessive use, perhaps the reader won't be able to read behind it the feeling that this present has generated in me.*

Ever since my dissertation findings (Rimanoczy, 2010) about the motivations of traditional business leaders who became champions of sustainability, I have been exploring ways to develop such a mindset for sustainability. To do this I developed the Sustainability Mindset Principles as the 'scaffolding' to design learning goals and find appropriate activities (Rimanoczy, 2021).

I find the principles serve as a helpful guide, deconstructing a complex reality into smaller components that can be better embedded into the teaching. Some exercises intended to illuminate the principles are more powerful than others, and potentially address several principles at once. This was the case of the assignment that inspired the essays shared in this chapter, and which I have selected because of its simplicity and transformational power.

## The context

We live in a world of accelerating pace, where time may not be money anymore, but has become the precious resource needed just to get more things done. Technological advances in many industries now permit workers to accomplish more in a shorter time, with multitasking emerging as one result. We have increased our productivity by splitting our attention into different directions, and we have learned to react in more automatic ways to situations, with the apparent benefit that we cover more ground.

The downside to the pursuit of increased productivity is that we can become less thoughtful, and we are not fully present to what is happening since the attention we have available is more superficial. This is not only the reality of the boomers, but it is also of the generations that came after (Pînzaru, Vătămănescu, Mitan, Săvulescu, Vițelar, Noaghea & Bălan, 2016). Millennials are naturally faster at processing information than the older generation, but there is some evidence that their focus is more scattered. According to data from the Centers for Disease Control and Prevention in the USA, Attention Deficit Hyperactivity Disorder (ADHD) increased from 6% to over 10% between 1998 and 2015.[2]

At the same time, the seriousness of our environmental and social planetary challenges has become broadly evident, and we rarely encounter fully eco-illiterate individuals anymore. Everyone seems to have some first-hand experience of the problems we are facing: for some it is air or water quality, species extinction, weather-related impacts on their life, race or gender equality or access to health or education. The United Nations SDGs summarize our urgent tasks to shape a planet that works for all.

The uncomfortable truth is that ending unsustainability requires us to change almost everything we are used to: our way of consuming, producing, traveling and entertaining, to name a few. Such a magnitude of changes is certainly challenging. Some adaptive changes can occur prompted by external factors, and the unexpected global crisis wrought by Covid 19 is an example of how an external circumstance imposes changes in our behaviors that we would have never voluntarily adopted. Because Covid 19 is an external factor, the restrictions imposed on the population have encountered varying levels of resistance as many people have become impatient, and self-centered; they have questioned the cost/benefit of the sacrifices required, and, especially in the younger generation, have demonstrated their unwillingness to abide by clearly stated warnings by all sections of the medical and health professions.

But there is a less traumatic source of change, which is maybe more powerful: it is called intrinsic motivation. Since the development of the school of Positive Psychology, research has highlighted the differences between problem orientation (external focus) versus vision orientation (internal focus). Grant (2012) studied behavioral changes related to sustainability motivated by external factors such as threats, doom scenarios and survival, noticing that they didn't challenge existing mental models or shift thinking. On the other hand, intrinsic motivations supported by positive emotions like passion for a topic, social sensitivity, purpose or vision, have been noted to encourage creativity, engagement and energy for sustained efforts (Gagne & Deci, 2005; Grant & Berry, 2011).

To summarize, educators face a context of urgent planetary challenges calling for actions and solutions (problem orientation), but also demanding a more profound change in behaviors, a shift in our mindset that is grounded in intrinsic motivations. For this change to take place, however, individuals need to become aware of their own behaviors, values and purpose. And this cannot happen haphazardly.

The exercise that will be introduced in the next section invites individuals to pause, to notice their own pace and to reflect on self.

## Developing a sustainability mindset

The assignment presented has been used in courses and modules intent on developing a Sustainability Mindset with students and in faculty development. When referring to a Sustainability Mindset, the overarching goal is to develop a way of thinking and being that can find its expression in sustainability actions for the greater good (Kassel, Rimanoczy & Mitchell, 2016). The path to this mindset is through a broad understanding of the ecosystem and our contributions to the problems, through awakening of social sensitivity and an introspective focus on our personal values and higher self. The contents to develop a Sustainability Mindset are grouped into four areas: Ecological Worldview, Systems Perspective, Emotional Intelligence and Spiritual Intelligence, with each area addressing specific learning goals stated as principles (see Figure 1.1).

### *The assignment: an hour in nature*[3]

The assignment focuses on the following goals and their respective principles. The goals are to:

- Practice slowing down mental and physical activity, changing their automatic responses into thoughtful pauses in order to get in touch with more profound insights (Principle: Reflection)
- Notice one's personal speed and develop comfort in pausing and more patiently exploring issues for a greater understanding of complexity (Principle: Reflection)
- Experience the power of introspection (Principle: Self-Awareness)
- Have an experience in Nature (Principle: Oneness with Nature)
- Understand that we are one with Nature, that we are one species within many others, and that all Nature is within us (Principle: Oneness with Nature)
- Experience different contemplative practices to have the opportunity to find what works best for them (Principle: Mindfulness)
- Develop an appreciation of the power of contemplative practices and develop new habits that increase mindfulness (Principle: Mindfulness).

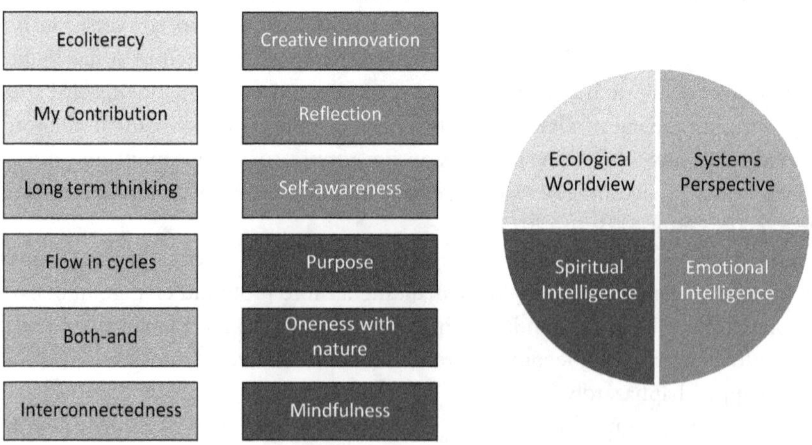


**FIGURE 1.1** The content areas and the 12 sustainability mindset principles, by author.

## *Timing*

This activity is best assigned mid-course, or towards the end of the course, since it is in contrast to more familiar cognitive and intellectual tasks. Assigning it later in the course gives the educator an opportunity to develop the relationship with the students, and to develop trust and a sense of safe space.

## *Positioning*

Anticipating the fact that some students may feel uncomfortable with this unfamiliar type of assignment, it is important for the educator to provide a clear rationale. This can be done in different ways:

- Sharing the learning goals, related to competencies like the ability to pause, notice and observe
- Connecting it to the development of key coaching and leadership skills by providing individuals with opportunities for introspection, reflection and developing self-awareness. Students may feel privileged to have an assignment they otherwise would not experience in the near future because these activities are mostly offered to higher management levels in corporations. These leadership competencies are key to their future success
- Referencing Emotional Intelligence as a key competency for their life, independent of their profession
- Using examples of corporations offering developmental perks to their employees such as mindfulness sessions or contemplative practices to help them find balance in their lives
- Offering it as an invigorating break from stressful times (most students feel the pressure and experience performance anxiety)
- Positioning it as a gift they are giving themselves, a rare occasion to officially "do nothing"
- Connecting it with creativity: we cannot be creative if we are busy focusing on efficiency and "getting things done"
- Finding ways to associate the assignment with the content of the subject that is being taught
- Announcing upfront that this assignment will be different and "unusual."

## *Instructions*

I have provided the instructions in writing and orally during our class, which gave the students a chance to ask clarifying questions, and gave me the assurance that they understood. Below is a text the reader can use or adapt.

> *Your assignment is to go out and spend one hour in Nature. Find a place that is relatively unfamiliar to you, meaning – avoid your backyard. Go alone, without a friend, without a pet. You can walk or sit. Don't bike, skate, and don't go roller-blading. Don't make it a sporting event: don't run, no paddling, boating, swimming or surfing. Don't take your phone with you. No book, no paper to write on, no music.*

> *Find a place to sit and just stay there. Do nothing. You don't have to meditate, however if your thoughts go to the things you "should" be doing; remind yourself that you are DOING an assignment – which is to sit and do nothing for one hour. THAT IS your assignment.*
>
> *There are no further instructions. When the hour is over (remember that you are not taking your phone), write your reflections about this experience. Write at least one page, and not more than two.*

## Adapting to different circumstances

I have used this exercise many times since 2010: during wintertime in New York, the temperature forced me to adapt to it. Instead of spending one hour in Nature, I asked them to spend one hour in a coffee shop, doing nothing. I kept the criteria of no cell phones, no music, no reading or writing material, and going alone. The reflections were equally profound. This said, the component of staying in Nature adds an extra layer of richness to the experience of "doing nothing" during one hour, disconnected from people, technology and the phones.

I also had to adapt it during the Covid 19 restrictions, where some students were not allowed to leave their home or drive to a park. Some had a park within walking distance, or did it at night when no police were around, or spent one hour alone in their own garden.

Some of the instructions evolved after seeing results that were less than ideal. For example, in one case a person went out to the backyard, and watched her dog running around. That focused the attention on the dog, and became a familiar dog-playing time. It was not the unfamiliar experience I was trying to prompt, and the reflections were shallow and insignificant. In another circumstance, an individual went out to the beach with his surfboard. It became an exhilarating time, yet it was something he usually does, so it was again not the unfamiliar experience I aimed at, and the reflections were superficial.

## Observations about results and impact

As the three reflective essays featured in this chapter show, there are some results we can anticipate, such as the experience of being without a phone. This is especially dramatic for the millennials. Every student commented on this experience in their essay. One student noted, "*this is the first time in my whole life that I was away from my phone for one full hour.*" Quite an experience, for sure and the most interesting insights coming from that "phone-less," time included increased self-awareness of one's emotional bond with the cell phone, with reflections such as dependency, addiction, lack of control, alienation from those sitting nearby and lack of attention to friends or family during meals and gatherings. The insights frequently extend towards the role of technology in their life, what it currently is and what they would like it to be. Several students become very self-critical, and want to make changes in their behavior, which is an indicator of intrinsic motivation.

Other recurring themes are the observation of their own pace, expressed through words like "impatience," "boredom," "wasting time," as well as feelings of anger or rebellion against the instructor who "dared" to impose such a situation on the students! However, in my experience there has not been a single case where that perception didn't change within the hour of the activity, giving space to insights related to a sense of peace, calm, relaxation, ease and quiet. The reflections that followed the experience are also valuable, and included taking a different perspective of one's personal "pace"; observing "oneself" from a distance; sharing intentions to slow down; creating pauses like this one and the intention of replicating the experience.

In many cases students report on thoughts connecting topics covered in class about nature, sustainability and values. This indicates that no assignment exists in a vacuum, but rather is reinforced by, and associated with, other activities or dialogues. The comments indicate experiential learning, as opposed to cognitive learning in classroom activities or readings. The assignment frequently elicits a more profound connection with Nature, a sense of oneness with, and an appreciation of, Nature as being more than a mere resource. It also surfaces guilt for one's personal behaviors towards Nature. In some cases, the individuals recall memories from their childhood, and the remembrance is seen in a different light, gaining new meaning, sometimes with a strong emotional content. Some individuals describe the experience of "getting back to a feeling they knew" and had not had for a long time. While I have felt uncomfortable at times for prompting emotional moments, I always checked with the students who have reassured me that while powerful and sometimes sad, it always was a valued experience, and they expressed gratitude for it.

New awareness of their contributions to the problems, or seeing things about themselves they don't quite like, is frequently followed by stated intentions to correct, repair, or change the situation. These may well be simply intentions, and I have no information about what effectively happens afterwards, but the first spark of an intention is something that may result in future transformations. Awareness is a one-way road with no going back – we cannot unsee what we have seen. We may look in a different direction, but our consciousness has already been expanded.

Beyond the common themes mentioned, the spectrum of topics brought up by this assignment is very wide. The depth of introspection, the interest in self-scrutiny and descriptions of emotions and feelings vary broadly and will depend on the individual's emotional intelligence, introspective habits and maturity. The capacity to express thoughts in words also will be a factor influencing the quality of the reports, whether due to self-awareness or to something technical like fluency in the language they have to write. Independently of these variations, what I noticed is that every individual doing this assignment has some kind of profound "aha" moment, which crosses pre-existing boundaries.

The observations are all anecdotal and empirical, and there is no current formal research design to capture, code and organize the material. It would be interesting to reconnect with the students after several months or a few years, to find out what if anything has changed for them as a result of this assignment. It may be difficult to isolate the impact of this single exercise, particularly in contexts where there are other assignments prompting self-discovery and reflection. But that fact should not diminish the merit of the power of such a simple and unusual exercise – to just be there for one hour, be with yourself and do nothing.

## Epilogue

I close this chapter with two paragraphs written in the essay after the Hour in Nature by a student. He wrote it in Spanish, and because of the poetic colors of his piece, I will share both the original and a translation.

> *He aquí que lo esencial, la belleza, es como la rosa que es sin porqué, florece porque florece, no importa si la miran o no. Nuestro cometido es ser, elegir, esperar y pretender lo bueno, lo mejor.*

*Vivir así en una época de intereses creados y mercancías, donde desterramos el dolor y soportamos el sufrimiento de vivir anestesiados.*

…

*Volver, volver, volver. El silencio habla siempre de volver a lo íntimo del ser. Reconocerse sujeto de tremendo amor y, por ende, vida maravillosa que contagiar. Asumir lo efímero de nuestro caminar, acoger la lógica del abrazo, más que la del contrato. Preferir servir. Preferir morir para vivir.*

*Therein lies the essential, beauty, like the rose that is a rose without a reason; it flowers because it flowers, regardless of whether meone is looking at it or not. Our purpose is to be, to choose, to wait – to hope, and expect the good, the best. To live like that in an era of vested interests and consumption where we negate the pain and tolerate the suffering of living as if under anesthesia.*

*To get back, get back, return. Silence always invites us to get back to our most inner self. To recognize yourself in a boundless love, and with a wonderful life to pass on others. To accept that our journey is ephemeral, embrace the hugs more than the love contracts. Choose to serve. Chose death so we can live.*

*Student 4, male, Spain*

## Notes

1 English is not the first language of these three students.
2 https://www.webmd.com/add-adhd/news/20181126/adhd-rising-in-the-us-but-why.
3 The inspiration for this exercise came to me from Hilary Bradbury, who a long time ago sent her students on a walk in nature. She has continued re-creating the idea of being in Nature as a catalyst for reflection and exploring our soul. Find more information here https://actionresearchplus.com/you-defend-what-you-feel-presencing-nature-as-experiential-knowing.

## References

Fredrickson, B. (2003). The value of positive emotions: The emerging science of positive psychology in coming to understand why it's good to feel good. *American Scientist*, 91, 330–335. 10.1511/2003.4.330.

Gagné, M., & Deci, E.L. (2005). Self-determination theory and work motivation. *Journal of Organizational Behavior, 26*(4), 331–362. 10.1002/job.322

Grant, A., & Berry, J. (2011). The necessity of others is the mother of invention: Intrinsic and prosocial motivations, perspective taking, and creativity. *Academy of Management Journal, 54*, 73–96. 10.5465/AMJ.2011.59215085.

Grant, G. (2012). Transforming sustainability. *The Journal of Corporate Citizenship, 46*, 123–137. Retrieved June 20, 2020, from www.jstor.org/stable/jcorpciti.46.123

Kassel, K., Rimanoczy, I., & Mitchell, S.F. (2016). The sustainability mindset: Connecting being, thinking, and doing in management education. *Academy of Management Proceedings, 2016*(1), 16659. Briarcliff Manor, NY: Academy of Management.

Pînzaru, F., Vătămănescu, E.M., Mitan, A., Săvulescu, R., Viţelar, A., Noaghea, C., & Bălan, M. (2016). Millennials at work: Investigating the specificity of generation Y versus other generations. *Management Dynamics in the Knowledge Economy, 4*(2), 173–192.

Rimanoczy, I. (2010). Business leaders committing to and fostering sustainability initiatives. Doctoral dissertation. Teachers College, Columbia University.

Rimanoczy, I. (2021). *The Sustainability Mindset Principles: A Guide to Developing a Mindset for a Better World*. New York: Routledge.

# 2

# HOW DOES EMOTIONAL INTELLIGENCE FEEL?

## An exercise to cultivate self-awareness, self-regulation, perspective-taking and empathy

*Lasse Lychnell*

### Introduction

As sustainability encompasses not only the planet but also people, emotional intelligence is considered to be an essential area of the sustainability mindset (Kassel, Rimanoczy & Mitchell, 2018). Working with a multitude of diverse stakeholders can bring up a lot of emotions, and the ability to manage these in a constructive way is imperative to solving global challenges. While teaching emotional intelligence has often focused on its conceptual parts, this chapter suggests how contemplative exercises might be used to provide students with a subjective and embodied experience of the phenomenon (Barbezat & Bush, 2013; Zajonc, 2009). Recognizing that emotional intelligence is a contested theoretical concept (Antonakis, Ashkanasy & Dasborough, 2009), this chapter focuses on a process that facilitates students' learning about their own self-awareness, self-regulation, and empathy, which can all be seen as important aspects of emotional intelligence (Goleman, 1998).

The chapter presents an exercise to cultivate emotional intelligence and is structured as follows. First, I provide a short background to emotional intelligence and how I use the concept in this chapter. Then I briefly explain the context in which the exercise was taught before providing step-by-step instructions. After this, I share anonymous quotes from students' reflection papers to illustrate immediate takeaways, as well as a letter from a student that illustrates some of the challenges and opportunities associated with teaching emotional intelligence at a business school.

### Emotional intelligence and the sustainability mindset

While emotion is an unconventional subject in a business school curriculum, interest in emotions and their management has increased in research and the popular press over the past decades. Metcalf and Benn (2013) suggest that emotion management is essential to dealing with and leading others through complexity. According to George (2000), emotions and moods play a central role in the leadership process, and emotional intelligence contributes to leader effectiveness in terms of, for example, developing flexibility in change and instilling

DOI: 10.4324/9781003229735-3

motivation in others. Compared to analytical and technical capabilities, emotional intelligence is further thought to play an increasingly important role at the highest levels in an organization (Goleman, 1998).

Goleman's (1998) popularized conceptualization of emotional intelligence includes five components: self-awareness, self-regulation, empathy, motivation and social skill. In leadership practice, emotional intelligence is not about being nice and being liked by everyone; rather, it involves giving difficult feedback and making tough decisions with equanimity and compassion (Goleman & Boyatzis, 2017). As sustainability issues involve a multitude of stakeholders with different interests, a lot of emotions are likely to be brought up. Turning such situations into opportunities for mutual understanding and collaboration requires the ability to manage emotions (Kassel et al., 2018).

Emotions may also be associated with a motivation to move toward sustainability. For example, a person who realizes a conflict between their espoused values and their values in action may experience feelings of guilt, sadness and anger. Rimanoczy and Sridaran (2018) suggest that such feelings play an important role in fueling action geared toward sustainability. Here, emotional intelligence may be helpful in using the feelings constructively rather than acting on instinct or just repressing them and acting as if nothing has happened.

George (2000) suggests that emotional intelligence consists of at least four aspects. First is the ability to appraise and express emotions – that is, to identify accurately how oneself and others feel and to find words to communicate this. Second is the use of emotions to enhance cognitive processes and decision-making. This concerns, for example, how attention may be directed to pressing concerns and helping decision-makers to choose among multiple alternatives. Third is an appreciation of the determinants and consequences of moods and emotions and how they evolve and change over time. This includes understanding how people will react in the face of tough decisions, as well as how one's mood may affect other people. Finally, the fourth aspect involves the management of one's own and others' emotions – for example, by exciting others and dampening one's own excitement when it will not contribute to the situation as a whole.

In sum, emotional intelligence focuses on the ability to be aware of, to articulate, and to regulate one's own and others' emotions, and to understand their consequences and act accordingly. This requires self-awareness, self-regulation and empathy. Developing these capabilities may contribute to a move toward sustainability and leadership effectiveness.

The next section introduces an exercise consisting of three parts and a follow-up reflection assignment. The aim of the exercise is to facilitate an embodied and subjective experience of emotional intelligence by:

- Increasing self-awareness and self-regulation by becoming aware of thoughts, feelings and bodily sensations in action without reacting to them
- Putting feelings and needs into words and communicating them to a partner
- Developing perspective-taking and empathy by taking another person's perspective into account and feeling what they feel
- Gaining a deeper understanding of why feelings arise in relation to universal human needs that are either satisfied or not satisfied.

In the next section, before providing step-by-step instructions for this exercise, I describe and explain the academic context in which it is used.

## The global challenges track and the "Being" course

I usually teach the exercise in question in a seminar on emotional intelligence at the Stockholm School of Economics within the Global Challenges Track, which is a mandatory part of the Bachelor of Science program in Business and Economics. The track is composed of four courses: "Knowing," which is about what we know about global challenges; "Doing," which focuses on what we are doing about them; "Being," which relates to questions like "Who am I in relation to the global challenges?," "What can I do and what do I want to do?" and "How can I lead for sustainability while at the same time being sustainable myself?" and "Expressing," which requires students to integrate their knowledge from the previous courses by carrying out a project that contributes to global goals. The cohort is about 240 students each year, and we teach the seminars in four groups of 60.

The intended learning outcomes of the "Being" course are threefold: (1) to account for and explain the meaning of key concepts, theories and different perspectives related to sustainable leadership, as well as how these relate to one another; (2) to learn and use a number of tools to reflect and develop knowledge about yourself as a leader; and (3) to integrate theory, practice and self-knowledge to discuss your own viewpoint on issues related to sustainable development critically. Participation and reflection make up 20 percent of the total examination. The students receive one credit out of 100 for each of the ten seminars they participate in, and another for completing a subsequent reflection paper. On average, about 95 percent of the students participate in the seminar on emotional intelligence and hand in the reflection paper.

The session on emotional intelligence follows a seminar on mindfulness, which lays a foundation by teaching the basics of attention, awareness, self-awareness, and their application in leadership (Goleman, 2013; Ricard, Lutz & Davidson, 2014). Before the seminar, I ask students to read one or two papers about emotional intelligence (for instance, Goleman, 1998; Kegan, Lahey, Fleming & Miller, 2014). The seminar involves two 45-minute sessions with a 15-minute break. To avoid an abstract and objective discussion, which may render the subject invisible, I prefer to devote as much time as possible to doing the exercise and then discussing theoretical questions in relation to the students' subjective experiences. Before the exercise, I usually begin by providing a brief foundation, outlining the major elements of emotional intelligence as defined by Goleman (1998), and ask the students what they think about the implications of emotional intelligence for business and sustainability, as well as their own careers.

## Exercise: "How does emotional intelligence feel?"

The purpose of the exercise is to put the concepts of emotional intelligence into action. It consists of three consecutive parts:

I. Exploring a situation that evoked strong feelings
II. Articulating and expressing feelings
III. Articulating and expressing needs.

The first part is an adaptation of hands-on practices suggested in the book *Search inside yourself* by Chade-Meng (2012), while the two latter parts are inspired by what I have learned through training in nonviolent communication (Larsson, 2020; Rosenberg, 2003). Each part consists of a number of steps, and one important task for the teacher is to make sure everyone follows. Making a timeline can be helpful. I usually allow three to four minutes for each conversation when I work with students in large groups (and longer when I work with executives in small groups). At the end of the seminar, I hand out a reflection assignment, which is described in the next section together with selected quotes from the hand-ins.

## *Part I: Exploring a situation*

The aim of Part I is to give the students an embodied experience of self-awareness, self-regulation and perspective-taking. This includes the ability to notice their own thoughts, feelings and physical sensations; to know how these render impulses to react and then to withhold these impulses; and finally, to shift from their own perspective to the perspective of another person.

### Step 1: Making pairs *and finding a situation*

- Ask students to sit in pairs and decide who will be Student A and Student B. A simple rule, such as Student A is the one with the longer hair, is sometimes useful.
- Ask them to think individually about an episode in their own lives that evoked strong feelings, either pleasant or unpleasant, that they are willing to share with their partner – for example, a situation at work, an episode in the family or an incident at school.
- Explain the importance of confidentiality and ask them to agree that what is shared in this seminar will not be shared with anyone else ever. This is important, because students need to feel safe. However, they also need to take responsibility for themselves and set boundaries for what they are comfortable with sharing. Students who feel uneasy with the exercise can choose a "safe" situation in which they experienced only positive feelings – for example, when they achieved a great goal. This approach also works. Ideally, the situation contains some degree of mystery or surprise that triggers their curiosity and willingness to explore it.
- Invite them to go back in time and relive the situation. You could start with a short meditation and then guide them with questions like, "What does the situation look like?," "Are there any particular sounds or smells?," "Who is there?," "What do people say and do?" and "How does it feel?"
- After a while, ask if anyone needs more time. If they do, give them another minute before inviting them to slowly let go of the situation and gently come back to the classroom and the present moment. Give them time and show them your respect as you do not know what they just have been through. Pay careful attention to the atmosphere in the classroom.

### *Step 2: Student A speaks, Student B listens*

- Instruct Student A to start sharing their episode while Student B listens and is present without interrupting in any way. (Interruptions include sympathizing, giving constructive advice and asking clarifying questions.) Student B should simply listen with curiosity, presence and empathy.

- Remind the students to stay focused on talking and listening while maintaining some awareness on what is going on within them – for example, impulses to interrupt, feelings of awkwardness when no one says anything, or feelings of discomfort when the listener does not approve of what the speaker says. Learning to pay attention to such subtle cues is key to developing self-awareness (Goleman, 2013). This is an excellent opportunity for the listener to explore self-awareness and self-regulation in practice.
- If the speaker runs out of words, instruct the pair to just sit in silence and wait until the time has finished. Usually, the speaker starts talking again after a while.
- Give them three minutes.

### *Step 3: Student B repeats what Student A said*

- Student B, who was listening, now repeats what they heard to Student A with the goal of making Student A feel totally understood. This could start with "What I heard you say was …" Emphasize sticking with what was heard and watching out for impulses to interpret what the other person said ("What really happened was …"), to sympathize ("I know how you feel"), to give advice ("I think you should …"), to encourage ("You did a great job"), or to talk about your own experience instead ("I remember when this happened to me …"). Maintaining mindfulness provides a good opportunity to observe some automatic communication patterns.
- When Student B has finished, Student A may point out if anything got distorted, added or forgotten. Then, Student B may complement their account by repeating these points. The goal is for Student A to feel completely understood.
- Give them three minutes, including the adjustments.

### *Step 4: Switching roles. Student B speaks, Student A listens*

- The students now switch roles, and it is Student B's turn to speak. Repeat the same instruction as before but with an important addition. Because the student who is going to listen now knows that they will have to repeat what the speaker says, this may easily be seen as a kind of performance. It is not. Advise students to forget about that and focus on remaining mindful, trusting that they will remember what is important when the moment comes.

### *Step 5: Student A repeats what Student B said*

- Student A, who was listening, now repeats what Student B said, as described in Step 3.

### *Step 6: Having a conversation about the experience*

- Give the students a moment to have a "normal" conversation about how they experienced the exercise.
- Remind them that you asked them to have some awareness of their own bodies and minds while talking and listening. Explain that it is now time to look at the thoughts, feelings, physical sensations and impulses that arose during the exercise. You may suggest a few questions, like "What happened?," "How was it for you?," "What was challenging?,"

"What was easy?," "What surprised you?," "What did you learn?" or "What insights did you have?"

- Give them three minutes for the conversation.

### *Step 7: Concluding the exercise and preparing for the next part*

- Ask if anyone wants to share with the whole group what they learned during the mindful conversation. Tell the students that they are not expected to share the content of their stories, but their insights and reflections on the exercise.

Besides inviting students to have an embodied experience of self-awareness, self-regulation, empathy and perspective-taking, the situation they shared with their partner provides the raw material for the next part of the exercise, which focuses on articulating and expressing feelings (George, 2000). Here is a good time to take a break before continuing the exploration with the same partnerships.

## *Part II: Articulating and communicating feelings*

While the focus of the previous part was on the episode as a whole, this part helps students more accurately to articulate and communicate their feelings. By doing so, the students also gain a deeper understanding of their partners' perspectives and an opportunity to feel how they feel.

Usually, we use a very limited vocabulary when we talk about feelings. For example, when I ask a student, or even a colleague, "How do you feel?" the response is generally "Good" (which is more of an evaluation than a feeling). In other words, while we may feel a lot, we rarely spend time or effort articulating and communicating these feelings. George (2000) suggests that accurately appraising emotions facilitates forming judgments and making decisions. Furthermore, labeling emotions is a way to facilitate mindfulness and self-awareness (Brown, Ryan & Creswell, 2007; Vago & Silbersweig, 2012).

### *Step 1: Putting feelings into words*

- Have students sit in the same pairs as in Part I and hand out a sheet with words describing feelings to each student. The sheet I use is inspired by and adapted from the Center for Nonviolent Communication.[1] It helps students to access a broader vocabulary.
- Instruct students to identify which words best resonate with what they felt in the episode they shared before the break – that is, the situation at work and so on.
- Many students quickly mark a lot of words, while others have difficulty finding a single one. After a while, ask them to choose the three words that best fit their strongest feelings. Some students think it is difficult to reduce the number of words, and it can be helpful to ask them if some of the words are variations on the same theme and if some seem more important than others when compared.
- Give them three minutes.

### *Step 2: Communicating feelings*

- Ask students to share with their partner the words they have chosen and to explain how these relate to the episode they previously shared with their partner.

- Tell them they have four minutes in total, and ask them to make sure both of them have time to share and explain.

### *Step 3: Discussing the exercise and sharing with the large group*

- Finish up by discussing with the partner how it was to put words to their feelings and share them with a partner.
- Finally, give students the opportunity to share with the large group any insights, comments or surprises.

## *Part III: Articulating and expressing needs*

The aim of Part III is for students to attain a deeper understanding of why the feelings identified in Part II arose by identifying what needs were either satisfied (and thus contributed to positive feelings) or not satisfied (consequently contributing to negative feelings) in the chosen situation. In this way, this part promotes increased awareness of oneself in general and one's needs in particular, and provides a glimpse into how other people function.

Rosenberg (2003), who provides a mediator's perspective, suggests that feelings arise in response to a need that is either satisfied or not satisfied. One may, for example, feel hunger when the need for food and nutrition is not met, or may feel encouraged when the need to be seen and heard is met. In this way, feelings may be a doorway to learning more about one's needs, thereby increasing self-awareness. To use business school jargon, this step can be seen carrying out a root cause analysis. This perspective is useful not only in meditation, but also in self-exploration.

### *Step 1: Articulating needs*

- Hand out a sheet with words pointing to universal human needs. Again, I use a sheet that is inspired by and adapted from the Center for Nonviolent Communication.[2]
- Ask students to mark which needs were either satisfied or not satisfied in the episode they shared before the break – that is, the situation at work and so on. Encourage students to raise their hand if they get stuck, because this step is usually more demanding than the previous.
- Ask students to finish with the three words that together provide the best explanation of the feelings previously identified. Sometimes, students find a perfect one-to-one match between feelings and needs; other times, the needs do not correspond to feelings as neatly. Here, it is more important that students remain true to themselves rather than providing a coherent but somewhat distorted analysis.
- Give them three minutes for this.

### *Step 2: Expressing needs*

- Ask students to share the identified needs with their partners and how these may have triggered the previously identified feelings.
- Tell them they have four minutes in total and ask them to make sure both of them have time to share and explain.

### *Step 3: Discussing the exercise and sharing with the large group*

- Finish by discussing with their partner how it was to put into words and communicate their needs.
- Create an opportunity for students to share in the large group any insights and reflections.

### *Step 4: Concluding the exercise as a whole*

End the exercise by asking the students to reflect on the seminar as a whole and take a moment to identify the most important thing they learned. Give students room for individual reflection and encourage them to make notes if that supports them. Then let them share with a new partner. Finally, let those who want to share with everyone in the large group. At the end, hand out the reflection assignment (described in the following section).

## Reflection assignment and students' learning

This reflection assignment should be handed out after the exercise to be submitted a few days later. It is aimed at consolidating what students have learned in the seminar, and asks them to go one step further by exploring a conflict they have experienced in their lives. The following instructions are given:

- Think about a situation in which you experienced a conflict (small or big), the outcome of which you were not satisfied with. Describe the situation as objectively as you can and how it ended.
- What feelings did you experience in the situation? What needs were satisfied or not satisfied?
- What feelings do you think the other person experienced in the situation? What needs were satisfied or not satisfied for that person?
- Can you see any alternatives that would satisfy both your own and the other person's needs? What would you say to the other person?
- What are your reflections after having completed the exercise?

In the following section, I share some of my students' anonymized reflections, which demonstrate an understanding of the seminar's core concepts and their implications. Many of the reflections refer to insights into the relationship between feelings and needs, as well as how this may be a path to a deeper understanding of oneself and others.

> *It's pretty cool that you can connect feelings to needs that are satisfied or unsatisfied. I have never thought of it that way before. I'm usually pretty good at understanding how other people feel, but to connect these feelings to needs has given me a greater understanding and insight, and perhaps a more sober and objective way to view situations and feelings. (Student 1)*
>
> *I think it is easy to "forget" the needs that lie behind others' actions. Sometimes, it's not enough to consider the other person's feelings. This is something I will take away from this exercise. (Student 2)*
>
> *This exercise has helped me realize how I can use my feelings to understand why I react as I do and how I can improve the situation. (Student 3)*

> *To be honest, I have previously had difficulties understanding people who get irritated or tired, because I'm a very energetic person. Therefore, it has been important to me to gain a deeper understanding that different people actually have different needs. (Student 4)*
>
> *I realize that the other person experienced the same resentment as I did, if not worse. One should try to feel what the other person feels and understand what's causing those feelings, even though it is difficult in the heat of the moment. (Student 5)*

Other reflections emphasize aspects related to self-awareness and self-regulation, such as being on "autopilot" and attending to the gap between stimulus and response in order to achieve greater flexibility in how to proceed.

> *I also realize that the concept of being on "autopilot" connects to emotional intelligence. The more often you're on "autopilot," the less adequate your emotional intelligence becomes. (Student 6)*
>
> *The reason why I was angry, disappointed, and stressed was that I, to a great extent, reacted immediately to what I thought was an insult. Taking two deep breaths would have made me see that the other person did not mean anything. It would have been more constructive if I could have scrutinized myself and, instead of letting my feelings rule me, would have felt empathy for the other person. Then, we would have had a much better chance to find common ground and find a way forward. (Student 7)*

For many students, there was a realization that different people may have very different perspectives, and it is important to understand these perspectives to find constructive solutions through communication.

> *I feel that I could have thought about the situation more objectively, and even though I don't agree with the other person, I realize what he means. It is essential to think about problems from many perspectives because my truth is perhaps not always the truth of everyone else. (Student 8)*
>
> *I have realized that one needs to see conflicts from both parties' perspectives. It may be difficult, but one really needs to think through how the other person may have comprehended the situation. Ideally, you have open communication because you don't have to guess what the other person is thinking. I understood the other person's perspective during the conflict, but even more so after finishing this exercise. It's because I had to think it through thoroughly and reflect on how both of us felt during the conflict. (Student 9)*
>
> *Put simply, most conflicts are possible to solve with good communication when one is prepared to see the situation from the other person's point of view. It is just so hard to remember that when you're irritated. (Student 10)*
>
> *I think it would have been easier for the other person to understand me if I had explained my situation in more detail, and then, suddenly, both our needs would have been satisfied. My last reflection is: OH MY GOD, how simple it would have been to avoid this situation. Things really don't need to be so complicated. One just needs to take time also for small things. (Student 11)*

Students further reflected on how this way of working empowered them to take responsibility for their own needs, take action themselves rather than blaming others for not understanding them and engage in open communication.

The exercise is a good reminder to stop when you feel something, think through what needs are not satisfied for yourself, and then act more efficiently to satisfy these needs and, at the same time, reflect on what the other person's needs are.

> *After the exercise, I realized the importance of understanding one's own needs, accepting that they are not always satisfied and that there are other ways to satisfy them than those you first considered. (Student 12)*

I wish that people, including myself, would be more mindful of the people around us/me and not let prejudices set expectations. If this were to happen to me again, I would have the courage to say no and explain my sincere thoughts and feelings.

> *I could have controlled my behavior better. Instead of being hungry and grumpy, I should have satisfied my need for nutrition by, at least, having a snack, as I know it influences my temper. By pondering if one should heat up because of such a small comment, one can avoid a dispute. I should have thought about the situation from the other person's perspective and that she was hurt. Instead, I just focused on her comment to me and magnified it to the extent that obscured everything else. (Student 13)*

Some students gained insights about themselves and how they act in conflicts, which revealed behavioral patterns they were unaware of before.

> *To express the feelings was also tricky because they may not really be what you want them to be. I was almost ashamed of admitting my reaction. (Student 14)*

> *I have gained an increased understanding of my difficulties and shortcomings in really seeing other people's perspective and the underlying causes. (Student 15)*

> *I actually take away quite a few reflections from the exercise. I rarely get into conflicts because I am afraid of them. I become very nice and agreeable, and I want everyone to leave the argument happy. It has always been that way. (Student 16)*

> *This reflection paper was interesting because I was trying to think about recent conflicts that I'd been in, and I realized that I rarely do confrontations. I get annoyed and upset, but I avoid uncomfortable situations as much as I can. Now that I've seen this pattern, I am going to acknowledge when it happens and think about if I would gain from a confrontation or not. (Student 17)*

> *Personally, I think writing these reflections is an excellent way for me to process a situation that was not entirely successful, and I genuinely feel that it will help me in similar cases in the future. It is something that I have found challenging my entire life. I'm sensitive and pretty impulsive, and to sit down and reflect has, thus far, been very rewarding for me. I am convinced that it will help me in future relations. (Student 18)*

> *This reflection exercise has forced me to think through a very painful but also important episode. I have reflected on how I behaved, but never really put myself in the shoes of the other person and asked myself what she must have felt during this period of her life. This insight gives me a new perspective on her life and my oversight of her needs. (Student 19)*

> *This exercise made me consider letting go of my pride and to get in touch with my friend again to talk. (Student 20)*

## Letter from a student

I want to finish this chapter by sharing a letter that I received from a student after the course had finished. It summarizes some of the challenges that come with teaching soft skills (such as mindfulness and emotional intelligence), as well as the great opportunities in terms of the changes in perspective and behavior that may occur. While this letter refers to the course as a whole, it centers on many of the aspects taught in the seminar on emotional intelligence.

> *I wrote this email a while ago but thought it was best to wait to send it until the final assignment was graded. I mostly want to thank you for the course and tell you how valuable it has been to me.*
>
> *At the beginning of the course, I have to admit that I was extremely skeptical about its content and what we were expected to do. Mindfulness felt like pseudoscience, and, to me, meditation was a rather uneventful way of killing time. The phenomena we discussed were not new to me either. Mental health and stress felt essential to talk about from a symbolic perspective, but it was nothing that would hit or affect me personally. On the contrary, all my life, I've looked up to people who worked hard, lacked work-life balance, and reached success at the expense of mental health. To some extent, I still do so; however, I can see that more clearly.*
>
> *Nevertheless, my definition of desirable qualities has changed. I have started to note these traits in other people with whom I'm interacting. I observe good listeners and controlling people. I also notice those who lose their self-control in discussions characterized by frustration and stubbornness of principle. That makes me reflect on how and who I want to be, and not to be, as a human being.*
>
> *My closest family has commented that, lately, I have become much more present and reasonable in discussions. I tell them about the gap between stimuli and response. When I get criticized, I try to practice open awareness, where I can interpret my feelings without judging them. When others speak or tell me something, I try to listen actively. It feels as if I have got new tools to handle more situations and my feelings. I like myself so much more now, and, for that, I am very grateful for this course.*
>
> *It feels like my linguistic ability is too limited to accurately express how revolutionary this has been to me, how clearly I experience the change within, and how strong the experience of change is when one is so convinced that you know best yourself. But this was my attempt. I will bring what I have learned in this course with me for the rest of my life.*

Thanks for that.

## Notes

1 https://www.cnvc.org/training/resource/feelings-inventory.
2 https://www.cnvc.org/training/resource/needs-inventory.

## References

Antonakis, J., Ashkanasy, N.M., & Dasborough, M.T. (2009). Does leadership need emotional intelligence? *The Leadership Quarterly*, *20*(2), 247–261.

Barbezat, D.P., & Bush, M. (2013). *Contemplative Practices in Higher Education: Powerful Methods to Transform Teaching and Learning*. San Francisco, CA: Jossey-Bass.

Brown, K.W., Ryan, R.M., & Creswell, J.D. (2007). Mindfulness: Theoretical foundations and evidence for its salutary effects. *Psychological Inquiry*, *18*(4), 211–237.

Chade-Meng, T. (2012). *Search Inside Yourself: The Unexpected Path to Achieving Success, Happiness (and World Peace)*. New York: Harper One.

George, J.M. (2000). Emotions and leadership: The role of emotional intelligence. *Human Relations, 53*(8), 1027–1055.

Goleman, D. (1998). What makes a leader? *Harvard Business Review, 76*(6), 93–102.

Goleman, D. (2013). The focused leader. *Harvard Business Review, 91*(12), 51–60.

Goleman, D., & Boyatzis, R. (2017). Emotional intelligence has 12 elements. Which do you need to work on? *Harvard Business Review, 84*(2), 1–5.

Kassel, K., Rimanoczy, I., & Mitchell, S.F. (2018). A sustainability mindset model for management education. In K. Kassel & I. Rimanoczy (Eds.), *Developing a Sustainability Mindset in Management Education* (pp. 3–37). New York: Routledge.

Kegan, R., Lahey, L., Fleming, A., & Miller, M. (2014). Making business personal. *Harvard Business Review, 92*(4), 44–52.

Larsson, L. (2020). *Nonviolent communication i praktiken 2.0 [Nonviolent Communication in Practice 2.0]*. Svensbyn: Friare Liv.

Metcalf, L., & Benn, S. (2013). Leadership for sustainability: An evolution of leadership ability. *Journal of Business Ethics, 112*(3), 369–384. doi:10.1007/s10551-012-1278-6

Ricard, M., Lutz, A., & Davidson, R.J. (2014). Mind of the meditator. *Scientific American, 311*(5), 38–45.

Rimanoczy, I., & Sridaran, K. (2018). Sustainability and the being dimension. In K. Kassel & I. Rimanoczy (Eds.), *Developing a Sustainability Mindset in Management Education* (pp. 3–37). London: Routledge.

Rosenberg, M.B. (2003). *Nonviolent Communication*. Encinitas, CA: Puddle Dancer Press.

Vago, D.R., & Silbersweig, D.A. (2012). Self-awareness, self-regulation, and self-transcendence (S-ART): A framework for understanding the neurobiological mechanisms of mindfulness. *Frontiers in Human Neuroscience, 6*, 296.

Zajonc, A. (2009). *Meditation as Contemplative Inquiry: When Knowing Becomes Love*. Barrington, MA: Lindisfarne Books.

# 3

# TRANSFORMING MINDSETS FOR RESILIENCE AND GROWTH

## A reflection of an educator and her (former) students

*Karthyeni Sridaran, Royina Bakshi and Rocky Reynaldo*

### Educator Karthyeni's reflections

In more recent years as I stare into the evening of my life, I have begun to reflect on my role as an educator more deeply. My research interest in Workplace Spirituality offered me the unique exploration of my own inner world and taught me to ask some profoundly deep questions. "Who am I?; Why do I do what I do?; Where am I heading in the bigger scheme of life?; What is my impact on my students?" are just a small sample of a laundry list of big life questions. I am well aware that I am part of an elite production line. As an academic, a career in mainstream academia meant a relentlessly busy schedule. The coveted tenure is undoubtedly a hallmark of success for surviving a stressful system designed primarily for the pursuit of the old maxim "publish or perish." Consequently, giving up on tenure aspirations meant I had to find meaning and be resilient in my strengths as a passionate educator. Just as I demanded my students in my class do a deep dive into their own selves, I was forced to go deeper too, confronting "the why" of my career. Interestingly, this reflective questioning takes time and, in all honesty, only recently did I gain solid clarity. The social labels slowly but surely start to loosen their grip, and I stand grateful for the time I spend with young minds, as an educator. It's a privilege to help shift mindsets in the classroom. The "progressive precariat" label (Standing, 2014) bestowed on those like me, no longer stings. The author is referring to a new class of the highly educated progressive group, who suffer a sense of alienation, of anxiety due to lack of security in their careers or jobs. The toss between earning a living and living meaningfully is humbling. He warns this class may be prone to anger – thus the title of his book "The New Dangerous Class." However, my spiritual practice and training over three decades helps me see that on the flipside of the raw emotion of anger is, in fact, the high octave cousin called passion and purpose. It is proof that we make the meanings in our lives. I had to shift my own mindset several times in my own life and the honesty of one's life experience is an innate part of an educator's suite of skills in the classroom. My power in my classrooms comes from this genuine desire to engage with students with honesty, empathy and a mission to contribute towards shaping young minds, so that they march forth living meaningful engaged lives, bringing solutions and prosperity to their own future social eco-systems.

DOI: 10.4324/9781003229735-4

In an education system that does not offer any time for deep inner reflection, neither the educator nor the educated ever pause to turn the lens inward for self-scrutiny. It requires taking time out to reflect deeply on our endless multidimensional selves. In the past two decades, I have taken annual time outs and maintained a regular practice of spiritual inquiry. I have barely scratched the surface of my own mind despite the regular dedicated time in exploring my innermost core. To find answers to deep important questions, and to confront, if I'm part of the solution or part of the problem? It allowed me to see clearly that my ego partook in the pride of being part of a so-called noble profession – buying into the marketing machinery when told to revel in our role of producing world-class future leaders. It is an enticing parable to adopt to inflate my own ego, until I take a peek out of the window of my lofty ivory tower into the real world of the educated class and see what a bloody mess it is! I concede – I am definitely part of the problem.

For over 20 years I have taught a variety of Management subjects in a variety of business schools to tens of thousands of students from around the world. In the brief moment that I encounter these brilliant minds in my classroom – I am mindful of the tiny window of opportunity to plant tiny seeds of growth mindsets. It is rare indeed to ever see their blossoming or even keep track of these past graduates who move on to the following chapters of their lives. In the face of mammoth global ecological and sustainability concerns, management education needs to produce clear thinking resilient leaders with an inclusive and a sustainable mindset to go beyond business as usual mentalities. I have learned from my students that when I bring my own humanity to class, displaying with some self-disclosure my own anxieties and vulnerabilities, it creates a safe place for students to share. Sometimes I get lucky and an authentic genuine atmosphere of inclusiveness is created. In the age of distractions, true engagement with students comes only when there is trust. Trust is tacitly felt and earned human to human. A reflexive model of learning allows students to be front and center and ask deep questions of themselves. Along with my former students, we share our experiences in a subject called Management Competencies where students were encouraged to become self-aware and given tools to do reflexive learning. In the following sections, my former students, Royina and Rocky, provide a rare glimpse from a student's perspective, their experience in my classrooms and their account of the inner mind-shifts that followed.

## (Former) student Royina's reflection

When I joined Karthyeni's class I had already made up my mind. About two years prior I had been diagnosed with Type 1 diabetes (T1D), an invisible chronic illness. As a result of this I had experienced earned stigma, as many people wrongly believed I got diabetes as a result of an unhealthy lifestyle, and ascribed stigma, due to negative associations with diabetes (Vickers, 1997). I had decided that people would stigmatize my condition and undermine me. I was reluctant to share any information about myself, or my condition, unless necessary. This was due to my self-perceived feelings of rejection, loneliness, interpersonal insecurity and loss of self-esteem that many with invisible chronic illnesses experience (Lambert & Lambert, 1979; Donoghue & Siegel, 1992, as cited in Vickers, 1997).

From the very first "Management Competencies" class, Karthyeni made it clear that this class would challenge how we thought and interacted within our own selves and with others. My first meeting with her was when I had to inform her of my condition. I expected her to simply acknowledge the condition and dismiss me. Instead, she asked me to explain all the basics of diabetes management to her. During this process she found out that I was living alone and offered to help out in case of any emergencies. I was taken aback by her empathetic response. As found by Nils and Rime (2012) empathy positively impacts relationships (as cited in Andreychik, 2019), and so, this simple act created a feeling of trust and acceptance for me.

One of the main activities in this class was to engage in a self-reflection at the beginning and then at the end of the semester. I chose to work on "Relationships and Emotional Intelligence." I kept a journal wherein I recorded instances where my EQ was challenged, how I felt, how I reacted and how I wished I had reacted. A pivotal moment for me, where I felt challenged, was when I had to disclose my condition to a project team that I was working with. Again, I feared rejection, dismissal and blame. Instead, disclosing my hidden area (Luft & Ingham, 1955) prompted my team members to be more open and share information about themselves with me due to the social principles of reciprocity and liking (Cialdini, 2001). They were also more willing to learn about the condition, which helped me feel like I moved from being a "Loveable Fool" who people would pity, to a "Loveable Star" who people liked and respected. Upon later reflection I realized that my anxiety and low self-esteem, related to how people might judge me due to my condition, was in fact stemming from my own personal insecurities. As Cooley (1902) explains, my self-idea consisted of how I believed others perceived me (which was someone to be pitied), my judgment of that belief (which was unfair) and my self-feeling (which was self-hate and mortification), I now understand to be my own faulty self-limiting belief.

Karthyeni advised us to take charge of our lives and actively shape our destiny, or learn to make peace with others making that decision for us. I decided to make a change when I realized how my lack of communication and stonewalling had been amplifying my own feelings of shame, low self-esteem, and rejection. I made a conscious decision to talk about my condition more openly. Social media can act as a tool to explore and challenge narratives of chronic illnesses (Johnson, Levings-Gregory, Jaclyn & Andrews, 2020) and I had previously leveraged this power through my YouTube channel where I shared videos to raise awareness about T1D in an Indian context. However, YouTube only allowed me to reach out to strangers and not people I knew. Inspired by other diabetics in the T1D community, I decided to use Instagram, where I interacted with a larger audience, as a tool to share more information about my condition. I believed that it would allow me to give people a more intimate insight into the mental and physical challenges of living with T1D. I started to consistently make posts about my condition, covering both good times and bad. I used different aspects of the four dimensions of speech, facts, relationships, self-disclosure, and appeal, to create an engaging yet informative narrative. Figures 3.1 and 3.2 are examples of subjects I posted about.

The posts about diabetes have an audience engagement rate of about 25 percent, as compared to my "normal" posts that is about 20 percent. People comment and message me saying that the posts have helped others with chronic illnesses feel less alone, and those without learn more about invisible conditions, and they have used my account

> tv_nerd14 I've had really bad blood sugars lately. I've been up 4-5 times a night, eating jelly beans even though I feel sick, because my blood sugars are going low. I wake up in the mornings with low blood sugar and I am drenched in sweat while my hands are shaking. Throughout the day, my blood sugars keep going really high, which means that I have to take multiple injections (can go up to 10) and avoid eating for hours. Sometimes this means that I only eat two meals in the whole day.
>
> As exhausting as this sounds the hardest part is having to fulfil other obligations like university, volunteer work, projects etc. as well as I would have otherwise. I used to try to explain how diabetes made this harder earlier but often that resulted in people either thinking that I'm making it up or taking away opportunities from me because I'm "sick". It's at these times that I wish I was "normal".
>
> There are so many hidden aspects of diabetes that are mentally,physically, and emotionally draining. Sometimes being a diabetic is really, really hard.

**FIGURE 3.1** Royina's social media post raising awareness of the physical and psychological challenges of living with a chronic illness

to help raise awareness in their own communities. Though there are still times when I struggle with living with a chronic illness, I do not feel as nihilistic as I did. I do not see it as a part of my identity anymore, but am able to see it as a part of my life instead. This allows me to broaden how I define myself and has enabled me to take on more and grow both personally and professionally.

I do not know if I would have achieved the same self-growth and confidence without the activity on introspection. Looking back on how I thought and felt about myself, and having the opportunity to challenge my beliefs about how people perceived me, has helped me take charge of my life and look towards making a change. I believe that I relate more positively with other people and my community now. I am grateful for the opportunity to grow. Certainly not every person in the class engaged at the same level of deep self-reflection as I did. My perspective is that, students may engage better if they are given opportunities to reflect on a pivotal life moment that shaped their present self-based on their lived experiences. Management Competencies has provided

**tv_nerd14** Since every single day is pretty much the same now, I actually forgot my third diaversary on the 11th!

It has been a very interesting 3 years with diabetes. When I was diagnosed, I was really depressed and often suicidal. I felt like my quality of life was very poor. A while ago though, someone said to me that if they had diabetes like me they'd want to die because my quality of life seems poor and I was baffled!

I have come a long way in the last 3 years. Now when people ask me what I feel about my diabetes, I actually say that it's one of the best things that happened to me. It allowed me to meet some of the kindest people in the world, it allowed me to feel like a part of a large positive community, and it serves as a personal reminder to make myself happy everyday because life can be really short.

Here's to being a Level 3 Diabetic and celebrating with some cake!

(P.S. this picture is from last December when we could actually eat at a restaurant)

**FIGURE 3.2** Royina's social media post looking back at her journey and growth as a diabetic three years after diagnosis

me with tools that shifted me from feeling helpless to a growth mindset, enabling me to see my health challenges as a blessing in disguise, giving me much hope for the future. I was fortunate that at the start of the class, the brief moment of self-disclosure with Karthyeni about my health care helped me develop trust and forge a meaningful connection.

## (Former) student Rocky's reflection

It was early August of 2019 during the second semester of my Masters study program when I first met Karthyeni as she was my lecturer for the subject called Management Competencies. I remember she introduced the subject as being different to any other subjects in the business curriculum.

"The focus is not on the business problem that you need to solve; here, the focus is on each and every one of you," (I am paraphrasing her pitch to the class).

"I am listening," I said to myself. I thought this subject is indeed unique and so I set an intention to listen and absorb the lesson.

The second week came, and it was about self-improvement. Karthyeni's words provoked me when she dissected the word "responsibility" by defining it as the ability to respond to a situation I am presented with. She then brought us to a rather intrusive dichotomy, i.e. I am either part of the problem or part of the solution – intrusive because on one side I am associated with sustaining a problem yet on the other side I need to take actions, hence requires effort to leave behind my comfort zone that is my inaction.

The dichotomy resonated profoundly with me as at the time I had been engaging myself in an unhealthy self-talk (Andersen, 2016) – I despised the notion of power, authority and politics to the point where I could well be sabotaging myself from leadership opportunities. I realized that my own negative perception on leadership had shaped my own (negative) self-esteem and self-concept.

I have always been a sensitive person, even as a young child. Perhaps, my sexuality had something to do with it. I often felt lost and could not fit in with the heteronormative society where a boy should prefer blue to pink, play with cars rather than animal dolls and idolize strong male action figures like Iron Man – on the contrary, I was more drawn to strong women figures, divas, such as Whitney Houston and Mariah Carey. I also became highly sensitive, as I faced various kinds of discrimination e.g. name-calling and microaggressions due to my Chinese descent and appearance. Adding to this complexity is my own negative view of authority. Having witnessed in my home-country and elsewhere, high-powered famous figures, usually men, get away with corruption and other perverse behaviors, I was more negatively impacted by these impressions than I cared to admit. I consequently internalized those early experiences by building a negative association with the idea of leadership – hence, affecting the way I behaved, i.e. refraining myself from participating in leadership opportunities. Karthyeni then brought an example of how sometimes people let negative comments from their past or their childhood define who they are today.

Something about what Karthyeni said not only made me realize more clearly these biases inside me, but also challenged my internal narrative as a victim of discrimination (Andersen, 2016) and made me take a step back and ask a different question:

"What if I am just afraid to be a leader because I am afraid to fail? Or worse still, what if I turn out to be like those leaders I hate?" This is what Pfeffer (2010) referred to as maintaining a positive self-image by simply not trying things out, therefore not bearing the risk of failure which may lead to a negative self-image.

As our first reflective exercise asked us to focus on competencies that we want to develop, I thought that I would use my newly found clarity and determination to take charge of my own destiny by improving my relationship with power and politics. At the time I was an Event Manager at Melbourne Business School Student Association (MBSSA). I decided that I would aim for a leadership role for the next semester. Ask and you shall receive – I became the President of MBSSA.

With this new awareness gained from reflective exercises in class, I found a new clarity and determination to take charge of my own destiny and be in the driver's seat of my own life and aim for a leadership role. Soon, I found myself at the helm of a coveted role as the President of the Melbourne Business School Student Association (MBSSA). Ask and you shall receive – it was clear to me that I had made an inner shift at this point.

My mission was to be my own change by demonstrating the qualities of the good leader I aspired to be. I was attempting to practice qualities from two theoretical leadership models, i.e. transformational and servant leadership which include creating a shared vision, building an inclusive community and encouraging innovation through service-oriented mindset (Smith, Montagno & Kuzmenko, 2004).

As a leader, I led by example, championing honesty, authenticity, even putting my vulnerable self on display as I shared my experiences of being disenfranchised in many instances due to my minority status. By doing so, I tried to connect with my team on a more personal level and at the same time instill the vision of my leadership that is to make students feel like they belong. In acknowledging that there is an alcohol drinking culture in Australia, which may not necessarily represent the many international cultures of our foreign students, my team and I instituted a change to appeal to non-drinkers. In the past, there was the assumption that all social events should involve drinking and as such were advertised to appeal mostly to those who drink. Additionally, I also ensured that regardless of the position within the committee, each member's voice was heard. It set the tone of the collaborative team dynamic I would like MBSSA to espouse.

**FIGURE 3.3** Author Rocky in light-blue tie in center front alongside with other MBSSA committees Semester 1, 2020. Photograph by author.

Throughout my presidency, I learned how to harness and be more comfortable with different types of power such as legitimacy, resources and network (Barsoux & Bouquet, 2013). With the legitimacy attached to my position, I was invited by the faculty to deliver an opening speech on 24 February 2020 to welcome new students during the orientation week.

In developing my own resource power, I gave my team freedom to contribute their resources and connection as long as the goal of broadening our student reach is met. This led to one of the committee members securing a collaboration with an organization that supports LGBTQ+ professionals in the workplace. Although the pandemic and lockdown measures thwarted some of our planned events, I found that over time attending networking session became easier and more natural for me as I recognized that I was driven by a larger mission to expand MBSSA's external collaborations. My concern for the constituents I represented gave me further impetus to go further.

Even at the end of my tenure, I still benefit from the networks I maintained. My good relationship with faculty staffs brought new opportunities into my life. Most recently, I was invited as one of the facilitators for the university's first Business Innovation Lab, in which I led five international students in a three-day design thinking challenge. For me, it was simply humbling to note how far I have come from my early days when I shied away from any leadership roles.

In hindsight, I realize that when we try our best to be a better version of ourselves, the universe somehow always finds a way to propel us forward by giving us the opportunity to work on ourselves. It took some inner shift to happen within me, resulting in all sorts of serendipitous encounters bringing just the right people at the right place and time. On reflection, Karthyeni's class was my turning point that moved me to a growth mindset. Since then, I feel like all kinds of wonderful opportunities came my way. It reinforced the lesson she often emphasized in class – that outer change begins with inner shifts. This whole experience reminds me of the lyrics from Bob Marley's Redemption Song in which he said that we need to emancipate ourselves from mental slavery for the only one who can free our mind is us.

It is unfortunate, however, that I saw some of my peers treating this subject only as another steppingstone to graduate without having a meaningful lesson from which they could benefit. I feel that young people in general have the tendency to build their own walls when confronted with information on how they should be resilient – they do not like something "preachy" because they feel like the speaker does not know anything about their life.

The feedback I have for institutions to have impactful lessons such as the one I experienced in this subject is to encourage course facilitators to set the tone in class by demonstrating that it is safe to show vulnerability. In genuinely sharing their experience of dealing with personal pain be it from failures, tragedies or misfortunes, facilitators minimize if not completely close whatever gap there might be between the students and them by creating a sense of relatability that comes from their life. Genuine sharing gives the instructor more credibility to speak about what could be a confronting topic such as self-improvement. Secondly, the facilitator could encourage students to write a similar reflection of their own and share what lessons they could draw from their own experiences. This might draw out the shy students to understand that resilience can be learned through practice.

## Educator's closing remarks

As the world changes at an unimaginable pace, most classroom lessons pressure students to keep up by cramming more and more. It may seem counter-intuitive, but perhaps the answer lays in teaching our students to slow-down, to reconnect with their inner-being. "The last thing a teacher needs to give her pupils is more information. They already have far too much of it. Instead people need the ability to tell the difference between what is important and what is unimportant" (Harari, 2018). It is every educator's role to provide students life skills, especially as they navigate an uncertain future, a good sense of self and to be resilient and willing to change and change again is the new norm forward. It was a great reward for me personally to watch my ex-students blossom before my eyes, a rare and cherished treat. I thank my co-authors, my former students. It was a privilege to collaborate on these reflections. As an educator, the transformation I witnessed in both Royina and Rocky has been rewarding and inspiring. I am positively hopeful that they will remain resilient individuals as they venture forth living their lives more meaningfully as valuable members of their respective social eco-systems.

## References

Andersen, E. (2016). Managing yourself: Learning to learn. *Harvard Business Review, 94*(3), 98–101.

Andreychik, M.R. (2019). I like that you feel my pain, but I love that you feel my joy: Empathy for a partner's negative versus positive emotions independently affect relationship quality. *Journal of Social and Personal Relationships, 36*(3), 834–854. https://doi.org/10.1177/0265407517746518

Barsoux, J-L., & Bouquet, C. (2013). How to overcome a power deficit. *MIT Sloan Management Review, 54*(4), 44–53.

Cialdini, R. (2001). *6 Principles of Persuasion*. Tempe, AZ: Arizona State University, eBrand Media Publication.

Cooley, C.H. (1902). *Human Nature and the Social Order* (pp. 175–185). New York: Scribner's.

Donoghue, P.J., & Siegel, M.E. (1992). *Sick and Tired of Feeling Sick and Tired: Living with Invisible Chronic Illness*. New York: W.W. Norton & Company.

Harari, Y.N. (2018). *21 Lessons for the 21st Century*. New York: Spiegel & Grau.

Johnson, D.J., Levings-Gregory, M., Jaclyn, C.P., & Andrews, L. (2020). Using social media to change the narrative around chronic illness. *The Australian and New Zealand Journal of Family Therapy, 41*(1), 67–79. http://dx.doi.org/10.1002/anzf.1400

Lambert, V.A., & Lambert, C.E. (1979). *The Impact of Physical Illness and Related Mental Health Concepts*. Englewood, NJ: Prentice Hall.

Luft, J. & Ingham, F. (1955). The Johari Window, a graphic model of interpersonal awareness. *Proceedings of the Western Training Laboratory in Group Development*. Los Angeles: UCLA.

Nils, F. & Rimé, B. (2012). Beyond the myth of venting: Social sharing modes determine the benefits of emotional disclosure. *European Journal of Social Psychology*, 42, 672–681. https://doi.org/10.1002/ejsp.1880.

Pfeffer, J. (2010). Power play. *Harvard Business Review, 88*(7/8) 84–92.

Smith, B.N., Montagno, R.V., & Kuzmenko, T.N. (2004). Transformational and servant leadership: Content and contextual comparisons. *Journal of Leadership and Organizational Studies, 10*(4), 80–91.

Standing, G. (2011). *The Precariat: The New Dangerous Class.* London: Bloomsbury Academic.

Standing, G. (2014). *The Precariat Charter: From Denizens to Citizens.* London and New York, NY: Bloomsbury.

Vickers, M.H. (1997). Life at work with "invisible" chronic illness (ICI): The "unseen", unspoken, unrecognized dilemma of disclosure. *Journal of Workplace Learning, 9*(7), 240. https://doi.org/10.1108/13665629710190040

# 4

# THE QUANTUM SHIFT

## Teaching sustainability leadership critically and mindfully

*Julia Storberg-Walker and Deyang (Mike) Yu*

### Introduction

This chapter focuses on a doctoral-level leadership course and describes how the course offers an innovative way to cultivate a shift from cause and effect dualistic thinking to integrative thinking and emergence. In other words, from an "unsustainable" to a "sustainable" mindset. We call this change the *Quantum shift*. "We" are Deyang (Mike) Yu, a doctoral student from China who took the leadership course, and Julia Storberg-Walker, the instructor of the course. Together we offer this chapter to illuminate a more holistic perspective of the design/implementation and impact of the course.

Two sources inspired the creation of the curriculum: Rimanoczy (2013) *Big Bang Being: Developing the Sustainability Mindset*, and Tsao and Laszlo's (2019) *Quantum Leadership: A New Consciousness in Business*. The former (combined with online professional development from the Principles for Responsible Management Education (PRME) Working Group on Sustainability Mindset) provided the sustainability mindset conceptual framework and some of the principles undergirding the course. The latter provided interdisciplinary research justification for a new leadership model with the capability to drive the shift from separateness to a consciousness of connectedness. Tsao and Laszlo (2019) was selected as one of the required textbooks on the course based on its evidence-based leadership model and its deep, trans-disciplinary synthesis of philosophical, scientific and human-centered ideas relating to leadership for planetary flourishing. In addition, the book's focus on developing a consciousness of connectedness through contemplative practices mirrored my (Julia's) understanding of neuroscience, as well as my beliefs about contemplative practice as a gateway to deeper consciousness.

I (Julia) understand the Quantum shift to signify a re-orientation to self and the cosmos anchored in a new relationship of interdependency and entanglement. The Quantum shift means that one is accessing/touching/being infused by the Quantum field. Based on my personal experience (e.g. a mind shift from a consciousness of separateness and independence to a consciousness of connectedness and interdependence), along with the research presented about the shift in Tsao and Laszlo (2019), I know that the Quantum field exposes the limitations of cognition, rationality, mechanistic processes and metaphors and binary thinking.

DOI: 10.4324/9781003229735-5

Consequently, acknowledging the reality of the Quantum field fundamentally challenges the very foundations of higher education and research. It is clear that contemporary higher education typically operates from a reductionist, cause and effect paradigm for teaching, research and service. Experiencing the shift can change all that, and the Quantum field infuses all facets of teaching and learning – from administrative decisions about resources and room assignments, to formal and informal curricular design processes, to delivering content, to facilitating activities, etc. – ALL of these elements of teaching and learning are fundamentally changed.

As thus described, the Quantum shift is directly connected to many of the key ideas of the sustainability mindset, including the four sustainability mindset content areas (1) ecological worldview; (2) systems perspective; (3) emotional intelligence; and (4) spiritual intelligence (Rimanoczy, 2021). The Quantum shift is connected to the area of ecological worldview by calling for the need to relate to nature differently (e.g. "my contribution"). The shift is connected to the area of systems perspective's both/and thinking through a non-dual understanding of the unity of the cosmos. The shift is connected to the area of emotional intelligence through practices of reflection and the development of self-awareness. And finally, the shift is connected to the area of spiritual intelligence through purpose, mindfulness and unity with nature.

As mentioned above, this chapter describing the Quantum Shift and its use in a doctoral classroom is the product of a collaboration between two people – the instructor (Julia) and a student (Mike), and it begins rather typically, describing the intellectual foundations of the course and related cognitive/intellectual elements of teaching leadership. Then the chapter weaves two different first-person accounts (student and instructor), in order to illuminate the perhaps deeper and usually unseen aspects of teaching and learning leadership for the Quantum shift, with a focus on *consciousness of connectedness*. The two accounts follow the practice of reflexive autoethnography (Ellis & Bochner, 2000), whereby personal experiences illuminate the larger teaching and learning culture currently dominant in higher education. These first-person accounts include cognitive, intellectual, intuitive, emotional and embodied elements – just as what happens in real life, in real leadership situations. These accounts are presented in italics to represent the fluid nature of reflexive writing.

The chapter provides details about the leadership course – including course design principles, integrating contemplative practices into the doctoral classroom, details about the quantum assignment and student writings, to illuminate how the course fostered some degree of new Quantum shift in the students. The chapter concludes with suggestions for how to successfully develop new-genre leadership courses that not only teach leadership, but courses that can possibly catalyze students into new understandings of consciousness and sustainability.

Why is learning about this course important now? To me, I (Julia) believe that the time for fostering the Quantum shift is now, and we need more people teaching and learning about sustainability/planetary flourishing than ever before. It is even more important to deeply understand what students and instructors bring into the classroom because teaching and learning about leading for sustainability/planetary flourishing can fundamentally challenge taken-for-granted assumptions about the role of higher education in society and how doctoral classrooms should attend to only the neck up – e.g. intellect only. Consequently, a more substantive understanding of hopes, fears, aspirations, goals/purpose and level of openness to new ideas is needed. With this understanding, the alchemy of transformative learning and teaching has a greater potential for impacting both students and teachers.

## Curriculum design principles

The Leadership Studies doctoral course I (Julia) teach in the Department of Human and Organizational Learning at George Washington University has changed over time to address contemporary leadership issues including gender, race, intersectionality, and most recently neuroscience, multiple ways of knowing, quantum ideas and leadership for global sustainability/flourishing. The name of the program is the Executive Leadership Program; it is cohort-based, and students are mid- to upper-level management primarily in the DC metropolitan area. Students work in diverse contexts – government, military, corporate, not-for-profit, healthcare, education, etc. The cohort program contains a bit over two years of coursework and then dissertation. The ages of the students typically range from mid-30s to 60ish. We have noticed a trend downward in the past few years.

The latest evolution of the course has been guided by two design principles: (1) the adoption of *critical leadership education* (CLE) principles to guide answering "what is leadership" and "how should leadership be conceptualized" and (2) the values-based inclusion of *leadership for planetary flourishing* (e.g. the Quantum shift and sustainability mindset) to guide answering the "why should I lead" and "what is the purpose of leadership" questions. While the first design principle is the foundation of the curriculum (actually, the foundation of all that I do as an adult educator), this chapter will focus more on the second principle because of its relevance to sustainability/planetary flourishing.

Together, the two design principles – CLE and leadership for planetary flourishing – combine to make a powerful critique of the dominant leadership paradigms, practices and textbooks. Both design principles support the premise of the two foundational books used for content (Rimanoczy, 2013; Tsao & Laszlo, 2019), namely that leadership needs to be thought of differently in order to generate sustainability. In addition, the two design principles weave together an interesting web of ideas around power, purpose, equality/justice, diversity of stakeholders and leadership as a way of being rather than doing. This last element – leadership as being – opens the door to the inclusion of mindfulness/contemplative practices as legitimate content/practices for this leadership studies course. Indeed, the Quantum shift requires some type of dedicated practice in order to access the level of consciousness described by Tsao and Laszlo (2019).

The course teaches leadership critically (Collinson & Tourish, 2015). A critical pedagogy challenges student's deeply held views about what leadership is and should be, and asks students to step outside of their comfort zones to consider the values, assumptions and purpose of leadership. Critical leadership education (CLE) principles legitimize a focus on leadership for sustainability/flourishing because the principles attend to the *end goal* of leading (e.g. for sustainability/flourishing) rather than just on leadership as an agnostic/value-free process. As described by Collinson and Tourish (2015), "Critical leadership education in general challenges the suggestion that various leadership skills can be taught as if they are neutral vehicles for achieving unproblematic ends" (p. 591).

The course is also designed to challenge students to experience new connections between mind and body through various contemplative practices and challenged them to reframe their assumptions about what doctoral-level education is and should/could be. Typically, doctoral-level courses address the heads of students only (Tisdell & Riley, 2019). However, as new leadership research suggests, leading for sustainability/flourishing is a full body, mind, heart and spirit endeavor (Tsao & Laszlo, 2019). And teaching for a sustainability mindset includes emotional as well as spiritual intelligences (Rimanozcy, 2013). Consequently, the leadership course engaged students in contemplative practices, included one aesthetic activity, engaged students with video clips and included weekly discussions connecting leadership studies with controversial current events. Course assignments included a three-week contemplative practice with pre- and post-practice reflections; an annotated bibliography type of assignment, and a final paper requiring students to compare/contrast two leadership perspectives (quantum leadership theory and a form of complexity leadership theory (Uhl-Bien & Arena, 2018)).

## Contemplative practices in the doctoral classroom: Educator's perspective

Related to the second design principle, I have introduced a variety of mindfulness/contemplative practices and assignments into all of my courses, including this leadership studies doctoral course. These practices generated a wide array of responses from the doctoral students, ranging from profound to cursory. This section will describe some of these practices/assignments, share some student writing and feedback, and hopefully will illuminate the deep connection between these practices and the development of leadership for planetary flourishing capabilities.

> *Educator's Reflection: I felt so vulnerable and paradoxically powerful at the same time … Before this moment, classes in my program were directed at cognition and intellect alone, not whole body, mind and spirit. When I stood and moved my chair from around the table to sit directly in front of the class, I was breaking the security of separation and distance. Inside I felt jittery (how mindful is that???) and yet I knew that I couldn't* ***not*** *do this. My deepening into contemplative practices over the past few years had bubbled over into my educator role and I could no longer enact nor condone a separation between the two. And at the same time I was aware of my audience – mid- to upper-level professionals, including highly-ranked military officers and C-suite leaders. I had no idea how this would be received … and so I sat, no instructor's table separating me from my students, and took a couple deep breaths. I wasn't a teacher in that moment; I was a fellow human being just trying to make a difference; trying to make the world a better place, and I was relating to my students in a profoundly different way … I think there's a big vulnerability in that space but also an acceptance, a surrender. A paradox of strength and vulnerability; fear and courage. I was a professor no longer professing; I was a fellow human being inviting each unique adult to experience a new relationship to self, to their intellect and to each other. And I was making this "ask" in an academic classroom, a space where emotions and feelings and spiritual matters were traditionally not valued. It was freeing and subversive to open to this new space of teaching and learning …*

I didn't begin right away with a practice. Because of the backgrounds of the students in the room, my strategy was to use scientific evidence to persuade students of the value of contemplative practices before actually experiencing one. Specifically, I described evidence from neuroscience and biochemistry. That first time I probably spent much more time using hard science to convince students of the benefits of contemplative practices; now, while I do reinforce the science connection, it takes a much smaller role in the "play" of contemplation in my classrooms. So the first time I led a contemplative practice I spent a few minutes talking "science," and then I invited the students into the practice, acknowledging that if they did not want to that was OK and they could just sit quietly in their seats. I also asked that no one type on or use computers or cell phones. I asked them to settle into their chairs and if they felt comfortable, they could close or lower their eyes.

> *I peeked to see what the students were doing during those first silent moments. Most had moved their chairs back from their tables and were sitting quietly with eyes closed. A couple were just sitting quietly. I breathed, closed my eyes, and went into the practice. It was short; it is called the "chemical cocktail" and I learned it from a former doctoral student who has gone on to generate wonderful work in embodied leadership. After a few centering breaths, I asked the students to imagine a moment of exquisite joy – a baby's smile, a loved one's laugh, breathtaking beauty in nature, a dog's wagging tail, whatever comes to them in the moment – and I asked them to literally smile and breathe, smile and stay in that joy for about 20 seconds. I peeked again during the silence while I was smiling thinking of my dog Buddy running to greet me. I noticed the silence and how the energy in the room felt different; it felt more anchored. After about 20 seconds I then invited the students to imagine themselves growing larger; sitting up stronger in their chair, and I asked them to move their arms out to their sides to take up more space. I asked them to* ***take*** *their space, to* ***be*** *large, to* ***own*** *the space. This again lasted about 20 seconds. I heard a bit of rustle from one of the students who had decided not to "play," and that was OK. At the conclusion of the practice I invited the students to come back into the classroom, and I shared that the exercise likely generated two chemical reactions in their bodies: (1) oxytocin, the love hormone, is generated by joy; and (2) testosterone, the power hormone, is generated by owning/claiming space. The combination of the two means that you're likely feeling joyfully powerful in class – and what a wonderful ground for learning that is!*

This first contemplative practice happened over a year ago, and since then students (but perhaps not all of my faculty colleagues) have become used to my classes being opened with a contemplative practice. Most importantly, the students themselves are now helping to institutionalize contemplative practices into the doctoral program! As is tradition in the program, at the conclusion of their period of coursework, each cohort gives a gift – a statue, a sign, a picture, etc. as they embark on their dissertation journey. Just recently (August 2020), the class that experienced my first contemplative practice described above donated funds to create a dedicated contemplation room on campus. So while not all faculty might see the value or appropriateness of contemplative practices in the doctoral classroom, the students evidently are!

## The role of contemplative practice: Student's perspective

As described above, the first few contemplative practice sessions started with talking about neuroscience and biochemistry and gradually ideas from quantum leadership theory were added in order to legitimize the value of contemplative practice for learning, focus and leadership. Importantly, students shared that they needed this "scientific" justification before becoming open to considering the merits of combining contemplative practices with doctoral-level education. The impact of the discussions and contemplative practices on student learning and worldview of course varied. While some students ultimately were not moved or transformed, others, like Mike below, has experienced a profound shift:

> *M: As a first-year doctoral student, I began the program after having been educated for two decades under the classical Newtonian paradigm. I deeply believed that scientific methodology*

> *should be as it was until I learned about the quantum field and the role of contemplative practices to enhance leadership. I had read about quantum science in pieces years ago, and I knew there were arguments about relativity theory and quantum theory. However, I didn't expect quantum science could change my life. Quantum science is obscure and distant to the public. Yet it should be taught to everyone in my opinion after I've learned about it. It is not a high-end technology nor advanced science that is for scientists only; I think of it as a set of cognition tutorials that helps in discovering the essence of self and the universe. With its help, one can realize how individual consciousness and collective unconsciousness may interfere with one's perception and emotion.*
>
> *Merely months after I formally got to know quantum science for the first time, I perceived my mindset shift simply from reactions to daily events in life. The perception is from both the emotional level and logical thinking level. Because we usually choose to intentionally overlook what we see or sense that we cannot understand through scholarly rational and objective language, we have ignored a large part of the world in our daily life unintentionally. Once we adopt a different lens to see the world, it reveals a different façade of it to us. The quantum approach is one of the lenses that are available to us. Throughout the semester, I gradually accepted the claims of quantum science and its implications for real life.*

## *The quantum assignment*

One of the assignments, worth 20 percent of the course grade, invited the doctoral students to select a contemplative practice that resonated with them and to practice it daily for at least three weeks. A copy of the detailed assignment instructions is provided at the end of this chapter. We discuss the wide variety of contemplative practices available to students, particularly as described in Tsao and Laszlo (2019) as well as the tree of contemplative practices from the Center for Contemplative Mind in Society (Figure 4.1).

The students had some challenges at first figuring out what practice they would be able to integrate into their already full and busy lives. I didn't want this to be a heavy burden, so I often assumed the role of coach to help each student find her/his own unique solution to the assignment's challenge. The students could start and stop the assignment at any point during the semester as long as they submitted the reflective paper by the due date. During the weeks leading up to the due date, we would check in during class on their progress and students would share their challenges and successes. This continual reinforcement seemed to help the students maintain energy and focus on the assignment.

The reflective papers, on the whole, were profoundly moving to read. Mike shares his here:

> *From the start of the Spring semester, I have been doing Qigong continuously for several weeks and anticipated a significant change inside of me. In this reflection piece, I will narrate my states of mind at the beginning and end of this experience and how the change was perceived.*
>
> *It was a surprise to me for such an experience when I started the leadership class. When I read through the exercises list in the textbook, I didn't realize that we were taking something such as Yoga, Qigong and Meditation, which I treated as personal hobbies, into academic experiments. In the pre-exercise piece, I was curious about whether I would feel something different by doing Qigong, which had been criticized a lot in China as related to too many fraudulent cases during my childhood time. I read through the instruction paragraph in the book very carefully and realized*

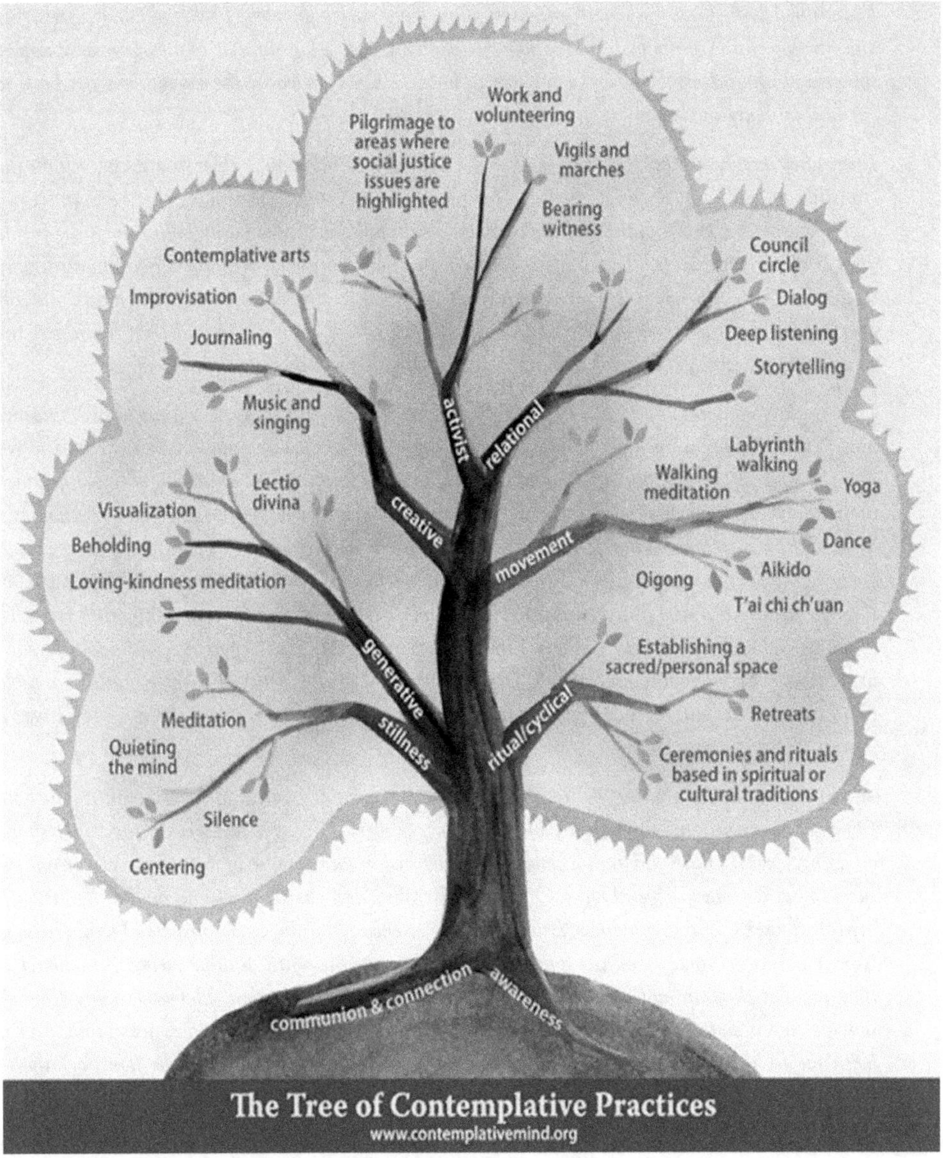


**FIGURE 4.1** The tree of contemplative practices. Used with permission.

*that other than the imaginary ball part, this instruction of Qigong is relatively substantial and solid, and then I started my exercise.*

*I used to try meditation with some video tutorials and reciting sutras, and I found those do help me with handling my mood and temper. At the beginning of doing Qigong, I felt the same as doing mediation or reciting sutras, I rapidly entered a mode of calm and peace. Then I began with the relaxation part, which asked the practitioner to let their body float, I did feel that my head was getting lightened in weight, although I was not sure if that was true or just my false perception. Then*

*I continued with the breath part, the most impressive part happened. I felt the energy gathering and moving inside of my body, not visible but clearly. It was a big surprise to me. After articulating the energy around my body for about five minutes, I slowly released the energy and got back to my normal body mode.*

*Even after merely a short ten-minute exercise, I felt my mind became clear and released from the stressful mode to some degree. It is not doing something to resolve the problems you encounter nor to let you forget about the stress for the moment as alcohol does, it magically built something inside your body to help you feel better about yourself. I cannot completely explain the feeling using my words, it is more like when you are working out a difficult mathematics problem for hours, and out of a sudden, you get your mind elevated to a different level of thinking and then figure out the solution to the problem. Doing Qigong to me is like the force that elevates my mind.*

*During the following days, I have been doing these steps every day before I went to bed, and my wife held her opinion on my "abnormal" behaviors. After about a week, I began to be able to get into the Qigong mode more rapidly and comfortably. It is so much like mastering a skill like playing badminton or boxing: you repeat the fundamental movements and practice and practice, and there will be a sudden change to you at a specific point in time. I can feel the imaginary ball of energy, or what the authors called "Qi" energy, more substantially. And every time I ran the "Qi" through my body, it was like a check of myself from another person's view, and my mind was able to see me, which was very strange to me in the beginning. With this experience, I also felt the connectedness between nature and myself. The interaction of energy between them is not like a cable but a field interaction. Even though I might sit alone and away from others, I felt deeply I was a part of the surroundings.*

*After the three weeks' practice, doing Qigong is one of my habits now. And I start to anticipate how Qigong is changing me through my mind, not my body. I didn't get stronger or into better shape by doing Qigong, but I did gain a sense of broader vision and a sense of duty to society or nature, speaking at the macro-level. Doing Qigong firstly calms and sharpens my mind, and then it built my consciousness of connectedness. Before such experience, obtaining consciousness of connectedness is more abstract. I understand the consciousness of connectedness just as other virtues, I know what they are, but they are not easily achieved as achieving a simple action. However, doing Qigong provides me a state of mind to realize how my consciousness of connectedness was built. And it helps me to be more concerned about others, not individuals specifically but the holistic "others" while making decisions.*

*From the quantum leadership perspective, such exercises are very helpful. Because the nature of each exercise differs pretty much, I am not very sure about how those exercises felt, but Qigong did bring me to a different state of mind. It helps me anticipate the consciousness of connectedness between me and the environment, and it moves my attention from focusing most on myself to a broader target. In general, such practices will help leaders to better understand themselves and others and to build a strong connection between them and others with a soft touch. They will enable the leaders to sharpen their adaptive skills by improving their openness and enhance their technical and emotional skills by establishing a sound foundation of connectedness.*

*Other than quantum leadership theory, we also read and discussed complexity leadership theory a lot. At the beginning of the semester, when I read both quantum leadership and complexity leadership, I felt complexity leadership sounded more familiar and more likely to work. However, after this exercise experience, I am much in favor of the quantum leadership theory. Complexity leadership theory provides clear views of different leadership types, and it worked perfectly for me as a tool to help me differentiate and analyze different leadership cases. But quantum leadership provides me with ways that I can practice to improve my ability to become a better leader, which is more effective for me.*

*In summary, this quantum experience broadened my vision of "quantum" field stuff and also leadership theories. I will continue such exercises and try different ones as well. To me, the most important part and the core of a good leader is responsibility, and quantum leadership motivates leaders to expand their responsibility from focusing on individuals and small teams to the bigger and broader environment, which I believe will make business better.*

## The content: Teaching and learning about leadership for sustainability/flourishing

In addition to the experiential component of the course, the strategy for teaching about the scholarly domain of leadership primarily involved developing the scholarly literacies for doing doctoral-level analysis and synthesis on two leadership models/theories: quantum leadership and complexity leadership. Both leadership perspectives consider leadership as distributed through the connections between people; leadership is not the behavior or skills or attributes of one person. However, the leadership perspectives varied considerably in terms of the underlying source of leadership. In quantum leadership, the source of leadership is the development of a *consciousness of connectedness*. In complexity leadership, the source of leadership is a *function* within organizations – it does not reside in a person.

Rimanoczy (20132013) highlights many of the diverse elements of both the quantum and the complexity view of leadership. For example, complexity leadership's history is from systems thinking; systems thinking is one of the key elements for developing a sustainability mindset. Likewise, quantum leadership is grounded in quantum theory findings illuminating the interconnectedness of all things and Rimanoczy (2013) affirms oneness with nature as a key principle of a sustainability mindset.

Class discussions explored the different dimensions of each leadership model and students "practiced" scholarly analysis and synthesis through an annotated bibliography assignment. The final paper called for students to compare and contrast complexity leadership with quantum leadership, and make an argument for one or the other as "better" equipped to address the planet's ecological crisis.

*Mike: What I learned about leadership?*

*A bit of background first: I have degrees in Economics and Finance, I worked for a bank and have experience with starting and running my own business. My purpose in enrolling in a doctoral program was rooted in my need to further develop my mind and my understanding of the world. Therefore, when I entered this course on leadership, I had paradoxical expectations, that is, I was longing for something groundbreaking but believed that there would not be a dramatic change in me merely from a single course. However, this course was my first experience of understanding leadership in a way that I believe portrays the reality of leadership most accurately. It emphasizes the essence of leadership and encourages a more human, intuitive perception of leadership. It is in contrast to what we usually do in social science, such as modeling and simplifying. I appreciate such an experience that opened a window for me to understand the world differently.*

*I didn't accept the quantum leadership approach in the beginning. In the first class, Dr. Julia guided the class in a short meditation. I barely did meditation before, or at least, I didn't expect to do something spiritual such as meditation in a social science class. I was confused, and I questioned why I am doing such an exercise in class in my mind. Ironically, I felt the difference such exercise made to myself afterward. Upon reflection, at that time I was not able to explain or comprehend my experiences with contemplative practice with my "scientific mindset." Now I think that this meditation exercise opened my eyes to things I used to classify in a category other than "academic" that may actually be scientific. I started to question the hard line between science and non-science.*

*In my previous leadership courses, the emphases were mainly on the development history of leadership theory and practical models that can be conveniently applied in real life. There were very few discussions about complexity and consciousness, and although sustainability has been discussed often, it was mentioned as a paradoxical problem. One common aspect of the theories and models I had learned is that they all have a bunch of assumptions, which simplified the scenario but lessen the practical range of the theories and models. I was used to this way of thinking until I started this course on Leadership. In this course, we were inquiring about complexities, dynamic movements, and activity chains with unknown patterns, which seem to be unsolvable. Nevertheless, after a semester of learning the new way, I realized that my previous cognition of academic study was limited and incomplete under the Newtonian paradigm. For years, I was chasing for the linear causality and effect mode of thinking; this course opened me to a new understanding that cause and effect thinking might merely not be applicable to all circumstances.*

*The quantum approach is destructive to the Newtonian paradigm. However, with the help of the Complex Adaptive System (CAS) approach (Holland, 2006), a logical connection can be made to help understand the complex causality, the reality of the object being studied, and the wholeness of the universe. It is difficult for a Newtonian scholar to accept the quantum approach as well as for students to believe what they tried to avoid. However, the Newtonian paradigm has long dominated academia and excludes others. I do not see it as the clash between science and pseudo-science, but rather I think the quantum approach is a new lens through which we can see the universe.*

*The quantum approach answered my question for years, that is, how can we balance between profitability and sustainability in business? It used to be a dilemma in a mindset that believes in individuation, the concept of time, and materialistic and atomistic vision, because, in such a paradigm, one is aiming to own the resource from the external environment and separate oneself from the collectiveness. It is true that if you maximize your profit, your descendent may not have the opportunity to do so in the future. Yet, if we step back to think about what this profit means in a different way, this dilemma might be worked out easily. And an alternative way is the greater purpose in the quantum leadership theory by Tsao and Laszlo (2019). With understanding the connectedness of consciousness, we are all of one wholeness, which means the perspective shifts from "me against you" to "me and you." If we can see each other in a relationship as organs of the same body, there will be much fewer controversies and conflicts.*

*There come the cons of the quantum approach. Due to its complexity and its nature, modern science is not capable of explaining and measuring all the aspects of the approach. Then, we are in a dilemma, which is we are not capable of measuring and matching the outcomes to the causality, and that will cause confusion between the true efficacy and pseudo-efficacy. As David Bohm explained in the concept of implicate order (1980), there is possible a different sort of connection of elements other than space and time, which means the separation might be derived from the deeper order. This challenge remains for further development in science to resolve. Nonetheless, I have learned that we should not reject something that lacks substantial evidence, and we may miss the beautiful scenery consequently.*

*In summary of leadership, I am confident that the quantum approach is a handy tool for leaders or leadership scholars. Because leadership studies focus on visions and measures, the quantum approach helps broaden the insight of a leader to the essence of his/her surrounding matters. The main challenge is to deal with the abstract paradigm shift and get used to the uncertainty and movement. Leadership theories under the Newtonian paradigm emphasize modeling and simplifying, which make the scene look better; however, without inquiring about the essence of the leadership and full range of details, we are losing the efficacy of our study. The quantum leadership theory turns out to be precise and valid especially under the increasingly complex era.*

## Conclusion: Leadership curriculum for catalyzing consciousness and sustainability

As illuminated by the course principles and first-person accounts of the instructor (Julia) and the student (Mike), teaching for catalyzing consciousness and sustainability requires (1) a reframing of the role of educator, and (2) a revising of the relationship between student and teacher. Regarding the role of educator: adopting a critical leadership education (CLE) perspective is a proactive choice that disrupts taken-for-granted norms and exposes the limitations of leadership as a series of skills, traits or behaviors. Instead, curriculum and pedagogy guided by CLE principles invite students to critique historical and contemporary leadership models and theories – and this critique opens their ability to see how the leadership industry (e.g. the publishers, consultants, bloggers, etc.) have been either unwilling or willing contributors to unsustainable business practices.

The relationship between student and teacher also changes from a "sage on the stage" to a "guide on the side" in many ways. When a teacher introduces contemplative practices into the curriculum, she/he is inviting this new relationship – a relationship of mutuality and co-creation. Yes, the course content needs to be delivered and grades need to be submitted. However, moving the relationship from just head to head (e.g. cognitive relationship only) to a fully embodied relationship (e.g. head, heart and spirit) means that the commonality of being human can be more salient during the semester, both inside and outside of class. This admittedly is a new space for many instructors, but one that I believe needs to happen because teaching for sustainability/flourishing is not about techniques or skills necessarily – it is about a state of being in relationship to others and the planet in a profoundly deep way. This relationship, or consciousness of connectedness (Tsao & Laszlo, 2019), must be experienced and practiced. It cannot be learned cognitively.

## Our closing contemplation: On connection, sustainability and flourishing

*Mike: From the course and contemplative practices, I perceived a significant change in my mind. It is noticeable when it comes to decision-making and critical thinking. Being a long-time student under the Newtonian paradigm has trained me with relatively static ways of thinking and learning. However, with the help of different "lenses," such as the quantum approach, I realized the way I know what I know may not be comprehensive, and the facts I believe may be temporary. Therefore, leadership as one of the core competencies in the modern world should be understood and educated through a mixed-format rather than being educated as a series of tools and concepts. The experience I had with Dr. Storberg-Walker is inspiring and educational to me, and I believe such an experience will be helpful to others as well. One takeaway of mine is that we need to soften the boundary of knowing. I have been educated in a way that emphasizes critical thinking but limits my knowing through strict frameworks. From the topics I have learned, a trend from uniform to complex can be perceived. I had been trying to find the most valid models, relevant assumptions and accurate responses. Nonetheless, I was pushed into a narrowed one-way that asks me to make trade-offs. Because of the narrowed insight, we have chosen many unsustainable practices in the business world. It is not because such practices dominate sustainable ones but the criteria of the decision-makers are limited to very short and narrow terms. If we could loosen the focus and embrace the complex, we may realize that there are hidden connections. A loss on a financial statement may not be a real loss for the business. With the appropriate angle to view the case and the appropriate stakeholders to consider, we will perceive the power of consciousness of connectedness.*

*On the other hand, from an individual perspective, I observed a cycle of learning in myself. At a young age, I was trying to absorb everything I saw. I imitate them. Later, once I had learned theories and principles, I started to think of everything I saw, with a reductionist view. I questioned my observations, and I tried to match them with the theoretical frameworks I knew. If I could not find a match, probably I will classify them as "unscientific" or way too "complex" and forget about them. The theories I have learned are all attempting to simplify the observed system, whether using assumptions or making estimates. With these theories, I tended to overlook what I do not understand, and I believed that there must be a simplified model for every complex system. I cannot say I was wrong, but after getting to know quantum science and achieving a little step in the Quantum shift, I realize that I will lose a large proportion of the world if I only focus on filtering and simplifying. Sometimes, I need to embrace the world as it is but not to find out a mathematical model that could describe a percentage of it. With this in mind, I could better understand the greater purpose and consciousness of connectedness in a way of logical thinking but not for the good of their names as I did before. A sustainability mindset is needed in the modern world either for planetary flourishing or even for the survival of humankind. However, many of us fell into a trap that taking sustainability as a morally correct word but not a real solution. We have seen businesses paying for sustainability campaigns yet making unsustainable but "profitable" decisions. If the financial statements look good, they will be satisfied. Nevertheless, the financial statements are just a model to capture the business wellness in a short term, such as a year, or even shorter as a quarter. We emphasize these models because they are intuitive and simple, but we overlook the fact that these models may not be capable to reflect the real business situation. Sustainability on the other side, as the word itself, is for one's own good. We need to see that, and a Quantum shift will help us to see that.*

*Leadership is about interactions between individuals and organizations, more importantly, it is about how the economy will move on. Once it comes to directions, it is necessary to maintain a long-term vision, which can also be called a sustainability mindset. Leadership is not only about skills, tactics and strategies, it is about relationships. With a mind of relationships between an individual and the others, the community, the society and the planet, or we can say with a mind of consciousness of connectedness, the world will be seen differently. Out of this course, I am very grateful for the Quantum shift, if I could call my experience one. I began to actively search and read in sustainability and transcendence literature, and I see myself differently in terms of leadership. I sense a drastic shift of focus from skills and tactics to connectedness and broad-mindedness. It is merely an initial step into a new world, but I can say it is a better world.*

### *For practice: Sample assignment from quantum leadership course*

Chapter 8 of the course text *Quantum Leadership* describes seven different practices that have been demonstrated (by research) to generate a sense of *connectedness* or sense of *purpose* in leaders.

This activity asks you to select one of the individual practices and "do" it for *at least* three weeks daily during the semester. *The activity will likely be more meaningful to you if you extend the practice for a longer period of time.* Note: the chapter includes two practices but they are not individual – Appreciative Inquiry and Group and Systems Level Practices – do not select these practices for this assignment. If you have an alternative practice to propose for this assignment, please discuss with me.

### *Steps for this assignment*

1. Select a practice and commit to it daily for a minimum of three weeks. There is no specific time requirement for the practice. Select one that is interesting to you and decide on a length of time that you can incorporate into your life without becoming over-burdened. Can you do ten minutes of meditation? Five minutes of yoga? Write one page in a journal? Be realistic and intentional. While progress rather than perfection is the goal, do look ahead to anticipate the best timeframe for completing the practice. Don't put it off till the end of the semester!
2. Prior to starting the practice, write a short reflective piece on the daily leadership challenges you are facing as a leader, or the daily leadership challenges you see in your workplace. Reflect and write on your leadership purpose, goals, aspirations. Keep this reflective piece for comparison. (two to four pages, double-spaced).
3. Do the practice for a minimum of three weeks.
4. At the end of the three weeks, and without looking at your earlier reflection piece, reflect again on the daily leadership challenges you have as a leader, or the daily leadership challenges you see in your workplace. Reflect and write on your leadership purpose, goals, aspirations.
5. Compare the two reflection pieces. Has anything changed? What has stayed the same? Did the practice contribute to your effectiveness as a leader? If yes, how? Did the practice have any impact in any area of your life? If yes, describe. Would you recommend this practice to leaders? Why or why not? Will you continue the practice? Why or why not? Will you consider a different practice at some point in time? If yes, describe.
6. Reflect on what you know about Quantum Leadership as compared to another leadership perspective of your choice from the course readings. Which leadership perspective do you believe would be more impactful/effective in your work context? Why?

7. Submit a short reflective paper, about three pages long, double-spaced. The paper should have a title page and page numbers, but because it is reflective APA guidelines are not required. The paper should include information from the pre- and post-practice reflections, a comparison of the two, what you learned from doing this exercise, and a short discussion addressing point 6 above.

## References

Bohm, D. (1980). *Wholeness and the Implicate Order*. London: Routledge.

Collinson, D., & Tourish, D. (2015). Teaching leadership critically: New directions for leadership pedagogy. *Academy of Management Learning & Education, 14*(4), 576–594. https://doi.org/10.5465/amle.2014.0079

Ellis, C., & Bochner, A.P. (2000). Autoethnography, personal narrative, reflexivity: Researcher as subject. In N.K. Denzin & Y.S. Lincoln (Eds.), *Handbook of Qualitative Research* (pp. 733–768). Thousand Oaks, CA: Sage.

Holland, J.H. (2006). Studying complex adaptive systems. *Journal of Systems Science and Complexity, 19*(1), 1–8. https://doi.org/10.1007/s11424-006-0001-z

Rimanoczy, I. (2013). *Big Bang Being: Developing the Sustainability Mindset*. London: Routledge.

Rimanoczy, I. (2021). *The Sustainability Mindset Principles: A Guide to Developing a Mindset for a Better World*. New York: Routledge.

Tisdell, E.J., & Riley, T.D. (2019). The landscape of mindfulness and meditation in adult education: A partial prescription (and critique) for lifelong learning and well-being. *New Directions for Adult and Continuing Education, 2019*(161), 9–20. https://doi.org/10.1002/ace.20307

Tsao, F., & Laszlo, C. (2019). *Quantum Leadership: New Consciousness in Business*. Stanford, CA: Stanford Business Books, an imprint of Stanford University Press.

Uhl-Bien, M., & Arena, M. (2018). Leadership for organizational adaptability: A theoretical synthesis and integrative framework. *The Leadership Quarterly, 29*(1), 89–104. https://doi.org/10.1016/j.leaqua.2017.12.009

Uhl-Bien, M., Marion, R., & Mckelvey, B. (2007). Complexity leadership theory: Shifting leadership from the industrial age to the knowledge era. *The Leadership Quarterly, 18*(4), 298–318. https://doi.org/10.1016/j.leaqua.2007.04.002

# 5

# EDUCATION FOR SUSTAINABILITY

## Sourcing inner qualities and capacities for transformation

*Christine Wamsler, Iris Maria Hertog and Lucia Di Paola*

### Introduction

This chapter is a reflexive case study of the Master's course on "Sustainability and Inner Transformation," and associated practices regarding its transformative potential. We begin by outlining the broader background and context that enabled the development of the course. We then go on to expand on its content, practices and evaluation. These aspects are also developed in detail by two participants, who describe the transformative experiences and mindset shifts that profoundly changed their way of seeing themselves, others and the world. The chapter concludes by offering some critical reflections on the lessons learned and ways forward for future work in education for sustainable development.

### Background and context

The "Sustainability and Inner Transformation" course is part of a Master's Program on "Environmental Studies and Sustainability Science" offered by Lund University Centre for Sustainability Studies (LUCSUS) in Sweden. It was formally launched in 2018, and runs annually over a period of three months. It was developed in the context of the Contemplative Sustainable Futures Program at LUCSUS. The program aims to critically assess the potential role of inner dimensions and transformation in societies' transition toward sustainability, and to create space and opportunities for knowledge development, learning and networking on the topic. It consists of three building blocks: research, networking and education. The course forms part of the education block.

The course was developed over a two-year period, which included experimenting with new approaches to learning for sustainability, their empirical validation and the creation of a better scientific foundation through targeted research on inner transformation and sustainability. Beginning in 2015, we worked in close cooperation with five students on the design of an Experimental Learning Lab. The aim was to explore contemplative approaches in sustainability teaching and learning, and assess their value and potential with a view to making them an integral part of the curriculum. The results showed that a majority of students welcomed

DOI: 10.4324/9781003229735-6

the integration of contemplative approaches into the Master's Program, and that they had a positive influence on their learning.[1] Scientific studies conducted under the program provided further support for course development, indicating the potential of inner dimensions and transformation for sustainability, and the achievement of the sustainable development goals (SDGs).[2] Overall, they showed that inner dimensions, inner transformation and sustainability were more connected than expected, in both research and education, with a potential that has not yet been tapped into.

## The course: Content, practices and evaluation

The overall purpose of the course is to develop a critical understanding of the potential role of inner dimensions and transformation for sustainability. Graded tasks include a written assignment and peer teaching. In the 2018–2019 academic year, written and oral evaluations (in both cases, response rates were 100 percent) and two surveys were conducted to assess the course's innovative approach and develop it further.

The course includes a series of lectures and seminars that are intended to explore the role of inner dimensions (i.e. values, beliefs, worldviews and associated cognitive, emotional and relational capacities) and their transformation to support sustainability. Issues such as environmental leadership, activism, social justice and human–nature connections are addressed. Knowledge from sustainability science, social neuroscience, psychology, behavioral economics, contemplative studies and transformation theories are systematically integrated in the form of lectures and seminars that also apply contemplative practices. The latter include self-reflection and visioning exercises drawn from various fields and sources, such as Joanna Macy's *The Work that Reconnects* (Macy & Brown, 2014). Self-reflection and enquiry are also supported by Weekly Councils and a Practice Lab.

The Practice Lab is intended to be an unconventional classroom component. It disrupts the traditional model of teaching and learning in the hope of offering something that is more transformative for both students and teachers. Its design was mainly based on lessons learned from the Experimental Learning Lab, which ran during 2015–2016. Compared to the first Lab, its focus is today more closely linked to the course's content (i.e. sustainability), and associated fields (e.g. climate change mitigation, climate change adaptation, sustainable consumption). Sessions build upon each other, moving from individual to social and environmental dimensions. Related practices are drawn from various sources, such as: (i) online tools recommended by the Lund University Student Health Centre, which provide exercises related to individual and social dimensions (e.g. breathing, body-scan, mindfulness, compassion, and loving-kindness exercises);[3] and (ii) the *Education for Sustainable Consumption through Mindfulness Manual*, which provides exercises related to the environmental dimension (Fritzsche et al., 2018).

In the course evaluations, all participants stated that the Lab added value to the course. The reasons were manifold. Students mentioned, for instance, that it had initiated or supported their own practice through providing time and introducing different techniques. This had, in turn, a beneficial effect on their overall wellbeing and learning processes. Regular practice was said to have led to increased concentration and improved self- and group reflection. Students noted: "I noticed that the days when we did the Practice Lab, my attention level in class was considerably higher," "I felt it was easier to focus in class, and at home on the readings" and it helped "Being more patient and thinking before speaking;

better listening." Students also stated that the Lab supported their learning environment, in that it influenced their relationships with others. They mentioned that "there was a different atmosphere in the classroom because of it." It also helped to deepen their knowledge through their own experience of how individual, social and environmental dimensions are interlinked. One student stated, for instance, that "our behavior is, to a large extent, influenced by our level of happiness, our fears and needs; these can all be influenced by mindfulness." Finally, the Lab was seen as important for aligning knowledge with action through "Practicing something that we were theoretically discussing" and "Walking the talk." Several students also explicitly expressed their appreciation of the critical mindfulness perspective, which was achieved by combining the Lab with lectures on critical mindfulness, wellbeing and stress, and their relation to sustainability issues such as consumption: "I have learned about the history, use and misuse of the concept. It has been contextualized and given broader relevance." Sixty-seven percent of participants also said that it had influenced their daily behavior, while the remainder were neutral. One of those who gave a neutral response stated that "When I did it regularly, yes [it influenced my daily life]."

Asked about which aspects of life the Lab had most influence on, the following answers were given: decreased stress/ anxiety (83.3%), increased attention during class and study (83.3%), increased overall wellbeing (66%), greater compassion and/ or understanding for myself and my work/ interests (83.3%), greater compassion and/ or understanding for others (66%), greater compassion for, or connectedness with, nature (50%), increased awareness of personal (un)sustainable consumption (50%), and increased awareness of others' (un)sustainable consumption (50%). The reasons given for lower, or no perceived influence were: (i) irregular personal practice, (ii) the short duration of the course and, thus, the Practice Lab (11 weeks, with seven weeks of in-class practice) and, consequently, (iii) the limited time dedicated to contextualized exercises (social and ecological dimensions). The importance of contextualizing contemplative practices in sustainability became apparent through statements, such as: "To me, the last few sessions in which mindfulness was put into a social and environmental context were the most relevant and valuable ones." This aspect also becomes obvious in students' accounts of transformation (see next section).

The post-course survey revealed that 79 percent of participants planned to continue their contemplative practices, and those who already had an established practice were planning to increase the time they dedicated to it. One student explained: "It makes me a more content and better person." Another noted that continuation is important to "Be the change you want to see," indicating its integrated value for both individual and global sustainability. Finally, all participants perceived mindfulness as being relevant in the context of sustainability teaching and learning.

In addition to the Practice Lab, each week ended with a Weekly Council. Here, the aim was to create a space for sharing theoretical and personal insights, perceptions, experiences and feelings in a way that allowed participants to be fully present, and learn from each other. Based on the council methodology (Zimmerman & Coyle, 2009), sessions were conducted in a circle with a focus on deep listening and understanding. Three guiding questions were defined to help structure the conversations:

1. WHY? Why is it important to consider inner dimensions in sustainability? (Rationales, related theories, concepts, disciplines/ fields)

2. WHAT? What is the potential influence of inner dimensions and transformation on sustainability outcomes at different scales (and vice versa)? (Rationales, theories, concepts, case studies, disciplines/ fields)
3. HOW? How do the identified rationales and influences translate into current sustainability research, practice, and teaching? (Concrete measures/ activities, related spheres of transformation).

All participants appreciated the fact that the Councils provided a safe place ("It was a safe space to discuss and increase knowledge") and offered an opportunity to: (i) listen to each other ("Listening to each other, and creating the time once a week to do that is excellent"); and (ii) reflect upon and systematize newly-gained knowledge ("The summing-up and open dialogue. Great"). The latter was seen as particularly important as the course took an innovative approach to sustainability ("Be able to express all those things that the course had set into motion"). It also gave students agency in their learning process ("It felt like we had ownership. This takes humility and trust in our abilities, thank you for that") and experience in walking the talk ("It was very useful to put contemplative practices into practice," "Using knowledge in our own practice, how we talk, listen, work together").

Overall, the evaluations showed that participants found the course to be very relevant. They highlighted its importance for personal and professional development, and the connections between them: "I am very grateful to have chosen this course. It combined personal development (leadership, self-knowledge and awareness, my own values and skills) with our academic choice (sustainability)." In a similar vein, another student stated: "Personally, it offered me hope for the first time since I started studying environmental studies and sustainability science. Professionally, it added another perspective to the common very limited, rational understanding of sustainability problems." The course was also perceived as transformative at both levels: "Inner transformation can offer both a straightforward scientific field within science, but also a personal practice that has the potential to improve both my personal and professional life." For some, this had immediate outcomes: "The course definitely changed me."

Another interesting result from the evaluation related to students' stress levels, which fell during the time the course ran (Table 5.1). Students described stress triggers related to: (i) their studies, in terms of pressure to succeed; (ii) the topic of their studies ("learning about the terrible environmental impacts we have on the planet and how unaware people are of them is always depressing and stressful"); (iii) their life context (often living for the first time in a foreign country and being in transition between education and work, with uncertainties regarding their future and meaning-making); and (iv) teaching methods (traditional methods were said to leave little room for deep learning[4] and self-reflection, which ultimately increased

**TABLE 5.1** Changes in perceived stress levels

| *Question: Do you consider the Master's Program and/ or your life situation as stressful?* | *Percentage of participants %* | |
|---|---|---|
| | *Pre-course survey* | *Post-course survey* |
| Yes | 21.4 | 18.2 (3.2% reduction in stress level) |
| Sometimes | 42.9 | 18.2 (24.7% reduction in stress level) |
| No | 35.7 | 54.6 (increase of 18.9%) |

stress levels). While actual percentages do not, as such, say much, qualitative comments from students who completed the post-course survey provide important information about the change in stress levels and how this relates to identified triggers. One student, whose stress level had reduced, noted, for instance:

> The reason why my answer changed so much is not that we have less to do than we had before (I still think that the amount of work in the Program overall is too much and this doesn't add to our learning experience – on the contrary I believe that it lowers the possibility to engage deeply with the topics). It is more that I gained a healthier perspective on the topic: I do as much as I can, but not [anymore] at the expense of my health and well-being.

Other given reasons related to an increase in positive emotions (motivation, hope), agency and meaning: "For the first time since I started studying I am motivated again and don't feel that we are just going to kill our planet … I learned super much about sustainability, but also about myself, and what I want to work on afterwards."

Given that teaching methods themselves were identified as a source of stress, the course's non-traditional approach also seems to have contributed to this change (Table 5.1). In this context, not only the Weekly Councils and the Practice Lab, but also the inter- and transdisciplinary approach ("It was amazing to have so many different perspectives on sustainability … truly an interdisciplinary course"), and peer teaching practices were mentioned. One student stated: "One of the parts I enjoyed most in the class was the peer teaching, because it narrowed down this huge and unfamiliar field to topics I and my classmates could relate to, it integrated the class into our own thinking process."

The fact that peer teaching was so successful is related to its positive impact on students' level of engagement and learning regarding the different topics they decided to look at. One assignment focused, for instance, on the value–action gap, i.e. the mismatch between knowledge and concerns about the environment, on the one hand, and the way people act (e.g. lifestyle or purchasing decisions), on the other hand (Hertog et al., 2019). The assignment was based on a survey (n = 97) sent to all current students and staff involved in the Master's Program, and a focus group discussion with everyone taking the Sustainability and Inner Transformation course. Interestingly, although participants only included people who were highly environmentally aware, their level of action varied from quite low (indicating a high value–action gap) to quite high (indicating the contrary). In addition, more than half of participants identified a value–action gap in their own behavior. Various inner dimensions were seen as potentially influential factors. Positive correlations were, for instance, found between pro-environmental behaviors, subjective wellbeing, and self-authoring mindsets (i.e., people able to critically evaluate and choose their own values, and determine their own path (Kegan & Lahey, 2009)). The survey sparked much interest among students and staff: "Why don't we talk more about this?," was a question that was commonly asked (Hertog et al., 2019: 35).

The personal and inner side of sustainability was a topic that many students felt strongly about and struggled with but is seldom touched upon in sustainability science and education. Reactions to the survey and focus group discussion showed that there was a clear need to engage with, and discuss, inner dimensions, the associated value–action gap and the inconsistencies and moral discomfort that comes with environmental awareness (Hertog et al., 2019).[5]

These observations are in line with the post-course survey, where participants said that they would like to see inner transformation theory and practices more firmly integrated into the curriculum and future educational activities. Moreover, 50 percent of students explicitly stated in the written course evaluation (and 93 percent in the oral course evaluation) that the Sustainability and Inner Transformation course was the best, or one of the best and most inspiring courses they had ever taken. It was described as "the missing piece in the [sustainability] puzzle," as it allowed students to link personal, practical and political spheres of transformation. [6]

## Students' voices on transformative learning

The previous section shows that the course allowed many students to experience transformative moments, which profoundly changed their way of seeing themselves, others and the world. When the course was run again in 2019–2020, students' responses were similar regarding its impact and their own transformation, illustrated by the following citations:

*"The course took a lot of 'space' in our heads – which is very positive."*
*"We were exposed to many new ideas."*
*"I very much appreciate the new perspectives the course brought."*
*"The course had a lot of provoking elements – I liked it."*
*"As intellectuals we should be open to explore new areas."*
*"Previous courses were too much on externality."*
*"For the first time in my sustainability education I felt as if I was 'coming home.'"*
*"The course helped me tighten up all the loose ends [regarding sustainability]."*
*"This course will be very memorable for me."*
*"This course is good for my life."*
*"I will take this course with me, into the future."*

That the experience of transformation is not momentary but can continue to influence students' lives on an ongoing basis is shown by the following accounts from two participants, Lucia Di Paola (Transformative account I) and Iris Maria Hertog (Transformative account II).[7]

### *Transformative account I*

The Inner Transformation course changed my life. It opened me up to a completely new way of thinking. It offered a new perspective for understanding sustainability. Up until that moment, our Master's Program had focused on the social, political, and cultural structures and systems of sustainability, and the natural science behind associated challenges. We largely focused on how to change such structures and systems to bring about positive change towards sustainability. The Inner Transformation course, in simple terms, taught us this: Structures and systems are made of people, and, if we ignore them, their inner values, beliefs, motivations and worldviews, sustainable change is not possible.

The focus on the individual was something completely new. It was not so much on the individual's life choices that impact sustainability (e.g. travel, diet consumption patterns), but on the values, beliefs, and worldviews behind those choices, which ultimately shape the political, social and cultural structures we live in. The course opened my eyes to the question: What and how can we change within ourselves that can help transform the current societal system towards sustainability?

In this context, mindfulness (one of many aspects explored in the course) was something that profoundly changed my way of seeing myself and the world. The course taught mindfulness in three different ways. The first was to present the most up-to-date theories on its benefits and pitfalls. Along with an exhaustive list of articles and books on the topic, we had the unique opportunity to hear perspectives from a diverse group of lecturers. Over a period of two months, we met sustainability researchers, practitioners, social entrepreneurs, psychologists, neuroscientists, and activists who work with mindfulness in different ways.

The second way we learned about mindfulness was the Practice Lab. At the beginning of each class, we had an optional ten-minute meditation session. Learning to keep your mind focused on the present moment is just like learning any other skill, like swimming or cooking. It is however almost impossible to learn to stay aware and let go of the continuous chatter of thoughts with a few meditation sessions. But, little-by-little, I improved and found more peace in my everyday life. Now, two years after the course, I have been meditating almost every day and, looking back, I realize what an impressive transformation I undertook.

Finally, mindfulness was taught and implemented in the management of the course. Both students and lecturers encouraged each other to create a non-judgmental safe space to exchange and learn. We felt comfortable expressing our emotions, we practiced active listening and non-violent communication, and some classes (e.g. the Councils and peer teaching sessions) had reversed "hierarchies"; students taught other students, and professors.

How meditation practice and mindfulness improved my life experience would fill many pages – I cannot stress enough how grateful I am for having had the opportunity to take the course. I am much more at ease with my thoughts and can, thus, manage and break negative thought patterns. Mindfulness, from my own experience, can have great therapeutic benefits for the individual. It can reduce negative emotions, stress, anxiety, depression and boost compassion, kindness and the experience of joy and happiness.

However, the course didn't aim to teach us how people can improve their lives for the sake of it. That would still be a noble cause, but perhaps not necessarily fit with our Master's program on Environmental Studies and Sustainability Science. The different course elements and learning approaches taught me that a more conscious and mindful community of individuals could be more prepared to support, embrace and welcome the emergence of a large-scale sustainable and just transition to a new society, a new system. The course pushed me to do more research and work in this direction, even after the end of my Master's studies.

I became involved in the "mindfulness for social change network," which was listed in our course database of institutions, stakeholders, networks, projects, and resources on the issue of inner dimensions and transformation in sustainability. Based on the understanding that the neoliberal, capitalist system is one of the main barriers to sustainability and justice, the question is if mindfulness could help us redefine our priorities, with intergenerational sustainability, justice (gender equality, education, health, access to the same opportunities, etc.) and love becoming more important principles than economic growth and profit accumulation. Related experience is seen in communities such as eco-villages and Transition Towns in Europe, which generally tend to see de-growth as a more fitting economic model and adopt mindfulness-based practices.

On a personal level, I have learned how mindfulness cannot only change my personal life for the better, it can also help me change the deep-seated structures and systems I live in. The Dalai Lama expressed this compellingly in the quote:

> When you have crises which are global by nature for instance, such as the environment or problems of modern economic structure, this calls for a coordinated and concerted effort among many people, with a sense of responsibility and commitment. This is more encompassing than an individual or personal issue … Of course, change must come from within the individual. But when you are seeking solutions to global problems, you need to be able to approach these problems from the standpoint of the individual as well as from the level of society at large. So, when you're talking about being flexible, about having a wider perspective and so on, this requires the ability to address problems from various levels: the individual level, the community level, and the global level.
>
> *(Dalai Lama XIV and Cutler 1998:166)*

### *Transformative account II*

After the first year of the Master's program on Environmental Studies and Sustainability Science, it was time to choose elective courses. There was a wide range of topics to choose from: water, landscape, gender, political ecology, social movements, etc. One of the courses seemed … different. It was called "Sustainability and Inner Transformation." My gut feeling said that this course could be either really good or really bad. On the one hand, it could lead to profound insights and finally connect the loose ends that my life seemed to consist of: science, sustainability, questions about what it means to be human, religion, creativity, love. On the other hand, it could turn out to be shallow, esoteric, or pseudoscientific. I took the risk, and I'm glad I did.

Today, nearly two years later, I don't exactly remember everything we did in the Inner Transformation course. But several experiences have stayed with me. One of them is an exercise that a lecturer from the Transition Network did with us (titled "Meeting the Descendants, or Future Beings"), which I later used myself in other contexts. In the first part, we were asked to walk randomly through the room in silence. At first, this was rather meditative, and we let our attention go to different parts of our body. Soon we were asked to speed up, and speed up, and speed up even more – imagining that we were in a hurry to catch a train. We were about 14 students running around, trying not to bump into each other. After a while, we slowed down again, and then stopped in front of a peer, a fellow human being. We were asked to make eye contact, and to realize that this person was just that, a fellow human being. One that stood together with us in the struggle for a more sustainable world.

In the second part of the exercise, we were asked to sit down on chairs placed in two, concentric circles. The chairs in the inner circle faced outwards, and those in the outer circle faced inwards, so that we were in pairs facing each other. The following text was read aloud by the guest lecturer (taken from Macy & Brown, 2014, slightly shortened):

> For those who are on the inside – you are just sitting here as yourself, just stay with your experience …
>
> For those in the outer circle, we are now going to do a piece of magic, of time travel. You may want to close your eyes … Imagine that the transition happens, that somehow

> we find a way to live sustainably on the planet, so that many years in the future there is a healthy, secure human presence here. Those on the outer circle, you represent people who live in that future time. Your people can look back on how humanity faced and overcame all the challenges, inventions, losses and wonders that made up the journey. And somehow you have developed the technology to travel back in time to meet your ancestors, those people who were part of the Transition, the Great Turning ... You have this extraordinary privilege to meet those people who helped to create the world in which you now live. The magic that allowed you people from the future to travel back doesn't allow you to speak, so you sit in silence. You have read about these people, perhaps seen films of them, or heard stories of that time. As you open your eyes, you get your first sight of this person who is from your past.
>
> For those in the inner circle, imagine that the person in front of you asks a first question – perhaps telepathically – and it goes something like this:
>
> The time you are living in is known in my day as the time of the Great Turning, the great transition, when humanity turned away from its path of continual growth and technical development and started to create more life based, sustainable ways of living. We know that at this time, there were still powerful corporations who put profit above care for the environment or for people; that governments were still slow to act. And, also, that there were small, yet committed groups finding new ways to live, and encouraging change. How was it to live back then, in 2018, at the time when these issues were becoming more urgent, when the response was just starting, how was it to live through these times?

I had no ready answer to such an all-encompassing question, but for the sake of this curious and silent person from the future, who was sitting in front of me, I started thinking aloud. Others around me did the same, filling the room with lively chatter. My words came slowly, wondering, followed windy roads. Most of the time I didn't know what I was trying to say and how to formulate my words, but sometimes I suddenly said something that felt true and valuable. And with this empathetic listener in front of me, giving me his full, silent attention amidst the chatter of my classmates, there was no need to rush. It felt like we searched together. After this first question, the "people from the future" were asked to move one chair to the right and, with a new person in front of me, came a new question. The guest lecturer continued, addressing the "people from the future":

> Now you have in front of you a second person living through this historic time, and a chance to ask a second question, which goes something like this:
>
> I imagine it is sometimes hard to see what was happening, or to find the courage and strength to act against the powerful forces. Or perhaps you are loving every minute of it? What supported you through those times, what helped to strengthen you, what kept you going?

Then, after trying to find an answer to this, came a new, silent, and curious future being, who asked the third and final question, spoken by the lecturer:

> When we look back at your time, we see how the actions of millions of individuals wove together to form the huge change that we call the Great Turning. What was your part

> in this movement, what was the thread you wove through that time? Perhaps today, you don't know that, but I'm interested to hear about what your passion is, what role you play, what your personal longing is? What are you wanting to change, or bring into being?

The questions touched me. Why do we talk so little about the things that really matter? But also: Do I even dare to believe that I am part of a bigger movement, that I have a role to play in this world, for real? Me too?

After these three questions, we switched roles. I became a listener. It felt like a privilege to be able to "ask" these significant and intimate questions to my classmates, and to witness their answers. I recognized much of my own struggles in their search for words. Gratitude, compassion, and respect welled up for all of them, all of us, simply doing our best to live a meaningful life in this complex and confusing world. None of us have answers, nor do we feel like we are changing the world, but we are all trying to build a loving and coherent life. Recently, I read three sentences that describe my experience so well: "You are never the one I believe you are. If I get to know you, I will discover that you are much more beautiful than I believed. And much more hurt" (Jean Vanier, quoted in Sjödin 2003).

It was a powerful experience, and one that I have wanted to share with others in different contexts. Loosely associated with the church that I'm a member of, I organize discussion evenings for those people who, for different reasons, do not feel at home in church, but still long to explore themes beyond a strict scientific-rational worldview. I call it "the Accidental Atheist," and we have been discussing themes such as longing, fear, meaning, love and grief. Each evening there is an hour of facilitator-led exercises (mixing physical and creative exercises with meditation, post-it-writing, and roleplay), and an hour of traditional conversation. This group has been a playground for me to create and test group exercises on different themes. This description makes it sound as if I know how to be a facilitator. Actually, I don't, but I followed the saying "Fake it till you make it," and it seemed to work.

When thinking about how to translate my own experience from the Inner Transformation course to this context, I realized that the exercise receives its strength from different elements. The first part is a good warmup. Walking together at different speeds seems to make people present in their bodies and in the group. Today, I use this in some form in almost every workshop. Secondly, there is something extremely powerful about listening to someone without being able to interrupt, and about just being listened to. No time travel simulation is needed for that, only clear instructions that the person who listens should do just that. And a question that is broad and deep enough to be able to set out on a windy road. Inspired by compassion meditation (which was a part of the Practice Lab), I've sometimes introduced such an exercise by trying to make people aware of the fact that they are "fellow human beings," in an attempt to recreate my own experience with listening and being listened to. For example, during an evening with the Accidental Atheist group about the meaning of life, I had people sit in pairs facing each other:

> To the people on side 1: From now on, you cannot talk anymore, you can only listen and observe. Look at the person in front of you, and try to feel compassion for them. Once upon a time, this person was born as a little child. And since that moment, they have tried to find their way in life. They are full of internal contradictions, but also long to experience a meaning with their life, just like you. And just like you, this person needs love to be able to live. The only thing you have to do in the coming minutes, is to listen to the person in front of you. Listen with curiosity and compassion.

> To the people on side 2: In front of you is a fellow human being who is here to listen to you, just like you are. They cannot speak, but are very curious to hear from you: When did you feel that your life was meaningful, that you were exactly where you were meant to be?

I used a similar setup for the peer teaching session at the end of the Inner Transformation course; here I let people talk about their personal experiences with the value–action gap.

Finally, there are, of course, dimensions of the exercise that are specific to the topic of a sustainability transition. I remember how hopeful it felt to allow the thought that everything might actually turn out well. Studying sustainability science means being confronted with climate change, biodiversity loss, habitat destruction, and various forms of human suffering. Add to that concepts such as lock-in, path dependency, power imbalances, societal structures and resistance to change. It is hard to have a positive outlook on the future. The first thing I think about in association with the year 2050 is doom scenarios related to climate change. But what if? What if we actually succeed in becoming a bit more human, a bit more rooted in love and care, and a bit less driven by the economic growth paradigm? Could it be true? We don't know. But maybe the point is that we need positive visions and hope to find courage and strength.

There were several lectures that made me shift my perspective. One I remember in particular. It described how our current society builds on a separatist worldview and, oftentimes, on functional relationships between people. It argued that we need another type of mindset for the Anthropocene; one that is relational, non-dualist and builds on resonant relationships. That is a lot to grasp, but somehow it brought together many different parts of my life. It also shed a new light on some of the experiences I had the week before. Afterwards, I wrote the lecturer an email:

> Somehow your lecture today brought together a lot of topics that I have been pondering about and that are important in my life … I might as well start at the end, with this fuzzy conception of "self" and "we" that we ended up discussing after the lecture. It's fascinating, I find. On the one hand it is so obvious, that I do not exist on my own, but that I have been deeply shaped by the people around me, by the languages that I speak and think in, by the science and culture that I'm born into. And also on a more practical level, that we are dependent on a large number of people all over the world for our food and clothes. But on the other hand, there is perhaps the fear of "losing myself" if I would start conceptualizing myself as fully "relational." Or maybe it is scary to admit how dependent I really am on other people. Being "strong" and "independent" might seem more safe (and more lonely). And then there is also the question of what that would mean practically, to live as a "relational being." What implications would this have? …
>
> It's funny actually – there is a part of me that can become skeptical if such non-dualist worldviews are discussed in a "scientific" context, but I have experienced them quite strongly in my personal life. Sometimes there is this realization that we as humans aren't so different after all, this moment of looking into somebody's eyes and recognizing yourself in them. Maybe this whole societal individualism is just an illusion we live in – like fingers on a hand might believe that they are really very distinct personalities, without realizing that they are a part of the same body …

> During the first meeting [of the Accidental Atheist group, themed around longing], we did some creative exercises, and then I invited people to write down on post-its what they long for in life. Seeing all the similar things that people wrote down (community, love, understanding), was really such a moment for me of realizing how similar we are as humans. It also made me wonder: if we all crave a life of more community and meaningful relationships, why do we still live in such an individualist society? How can multiple people who feel lonely live side by side and all believe they are alone in their loneliness? Even if it doesn't cause systemic change, we would all be happier if we would manage to have more "resonant" and less "functional" relationships!
>
> One big open question that I still have, is how that is to be applied to the human–nature relationship. You spoke about moving towards a less dualistic view on humans and nature, but what does that really mean? How can I get rid of these concepts in my head? It makes me think of another binary, the one between men and women. I have become close friends to a genderqueer person, and this allowed me (be it with some trouble) to create a third, genderless category in my mind. But, damn, it is hard to get rid of these categories, even if one makes a conscious effort to do so. And in order to move beyond the human-nature dichotomy, maybe we should get rid of the whole concept of "nature," no? It's hard to try to imagine what my bachelor program in "Forest and Nature conservation" would have looked like if we hadn't used the concept of "nature" … I see that there is something valuable in a non-dualist worldview, but I just honestly haven't been able to wrap my head around it yet.

I guess I still haven't fully "wrapped my head around it," two years later. But the journey continues. Since the Inner Transformation course, however vague it might be, the notion that my very personal feelings of hope and my longing for community and oneness are somehow connected to the struggle for sustainability, has grown dear to me. It offers me hope, even if I cannot clearly articulate why, in scientific terms. And, somehow, I trust that I will grow to understand more of how all these different things relate, if I only continue to engage with my head, heart, and hands.

## Conclusions

Research into education for sustainable development points to the need for a re-orientation towards more experimental, innovative and whole-person approaches that challenge the fundamental assumptions of mass education. This, in turn, requires the incorporation of new pedagogical models, where engaged students become agents for change in a learning community (Dawson & Oliveira, 2017). It requires learning that builds on experience and emotions that are embodied – and not only "embrained"; the latter are, therefore, better able to foster the behavioral change needed to act sustainably (Dawson & Oliveira, 2017). Despite recent advancements in the fields of contemplative and transformative pedagogy (Mezirow, 2000), putting these principles into practice is still a challenging endeavor.

This reflexive case study and the accounts from students who took the Sustainability and Inner Transformation course are a case in point. They show that embracing the above-mentioned principles is possible and, if realized, are beneficial and highly appreciated by students.

Ongoing empirical and theoretical studies are needed to improve (the integration of) emerging practices to source innate inner capacities and, simultaneously, solve problems and shift systems and cultural norms. The aim is to systematically support cognitive, emotional and relational capacities of being, knowing and acting with the most appropriate learning structures, processes and practices. In this endeavor, we also cooperate with others engaged in transformative education and practices that seek to explore diverse heuristics, methodologies and platforms for learning (Walsh et al., 2020a, b; Sharma, 2018).

The course, and the Contemplative Sustainable Futures Program, can be seen as part of an emerging nexus of integral education (Esbjörn-Hargens et al., 2010) and social change that is based on building a more conscious society. Others have called this "meta-modernism" (e.g. Björkman, 2018), or the interplay between systems, soul and society (Emerge, 2019; Perspectiva, 2019; Wilber, 1999).[8] As scholars, researchers and educators, we can choose how we position our work with respect to (neoliberal) institutional norms, growing worldwide resistance to these norms, and more integral approaches.

Inner dimensions and transformation for sustainability is clearly an underexplored field, and highly relevant in the context of education and the SDGs, particularly the global education goal (SDG 4) that aims to ensure that all learners are capable of contributing to sustainable development (Target 4.7). The notion of individuals as active agents of change towards sustainability is by no means widely accepted and, in fact, it conflicts with some dominant belief systems and worldviews (O'Brien & Hochachka, 2010). It represents, in and of itself, a fundamental challenge that calls for transformative learning and a questioning of values, beliefs and commonly held assumptions about human–environment relationships. Related approaches, however, are not always easy to pursue within the current research environment. Deliberate transformation, thus, also calls for deeper inquiry into current structures of meaning-making, not only within society, but within education and science itself.

## Notes

1 Further details regarding the Lab evaluation can be found in Wamsler et al., 2018.
2 The studies can be accessed from https://christinewamsler.wixsite.com/sustainable-futures/research-1.
3 E.g. www.headspace.com.
4 Deep learning is associated with a certain level of independence in choosing what is to be learned (Trigwell et al., 1999).
5 This statement should be seen against the background that sustainability education has, so far, focused on structural dimensions, the external world of ecosystems, wider socioeconomic structures, technology and governance dynamics, which students had studied for over a year. It highlights the need for more integral approaches that address political, practical and inner spheres of transformation – and not structural aspects in isolation.
6 Note that this section includes extracts from Wamsler (2020), which have been revised based on the authors' ongoing work; they were complemented with students' accounts to illustrate and analyse the transformative potential of the course and its teaching approach.
7 After her studies, Lucia Di Paola joined the Just Transition team at ICLEI Europe. ICLEI (Local Governments for Sustainability) is a global network of more than 1,750 local and regional governments committed to sustainable urban development. Iris Hertog started her career as a research assistant in sustainability science, before joining a Dutch consultancy that offers process management advice to businesses and governments related to sustainability transitions.
8 This emergent social change is seeking to integrate inner and outer dimensions and transformation, neither preferring one over the other, nor reducing one to the other, but finding ethical, skillful and effective ways to improve our capacity to foster sustainability research, education and practice.

## References

Björkman, T. (2018). *The World We Create.* Sweden: Metamoderna Publishing.

Dalai Lama, V.I.V., & Cutler, H.C. (1998). *The Art of Happiness: A Handbook for Living.* New York: Riverhead Books. Note: The indicated page refers to the Italian edition.

Dawson, J., & Oliviera, H. (2017). Bringing the classroom back to life. In *EarthEd (State of the World): Rethinking Education on a Changing Planet* (pp. 207–209). Washington, DC: Island Press.

Emerge (2019). Independent, non-profit media platform sowing the seeds of a new civilization, available at: www.whatisemerging.com/ (accessed March 30, 2019).

Esbjörn-Hargens, S., Reams, J., & Gunnlaugson, O. (Eds.). (2010). *Integral Education: New Directions for Higher Learning.* New York: Suny Press.

Fritzsche, J., Fischer, D., Böhme, T., & Grossman, P. (2018). *Education for Sustainable Consumption through Mindfulness; Active Methodology Toolkit 9.* Elverum, Norway: PERL-Partnership for Education and Research About Responsible Living. Available online at: http://achtsamkeit-und-konsum. de/wp -content/uploads/2018/06/Toolkit_EN_online-1. pdf.

Hertog, I., Betzler, L., & Di Paola, L. (2019). An exploration of the role of inner dimensions in explaining the value-action gap amongst sustainability science students and researchers. Final Paper. Sustainability and Inner Transformation Course, Lund University.

Kegan, R., Lahey, L.L. (2009). *Immunity to Change: How to Overcome It and Unlock Potential in Yourself and Your Organization.* Boston, MA: Harvard Business Press.

Macey, J., & Brown, M.Y. (2014). *Coming Back to Life: The Updated Guide to the Work That Reconnects.* Canada: New Society Publishers.

Mezirow, J. (ed.) (2000). *Learning as Transformation.* San Francisco, CA: Jossey-Bass.

O'Brien, K, & Hochachka, G. (2010). Integral adaptation to climate change. *Journal of Integral Theory and Practice, 5,* 89–102.

Perspectiva (2019). Research platform that seeks to build the intellectual foundations for a more conscious society, available at: www.systems-souls-society.com/ (accessed March 30, 2019).

Sharma, M. (2018). *Radical Transformational Leadership: Strategic Action for Change Agents.* Berkeley, CA: North Atlantic Books.

Sjödin, T. (2003). *Eftervärme.* Stockholm, Sweden: Verbum AB.

Trigwell, K., Prosser, M. & Waterhouse, F. (1999). Relations between teachers' approaches to teaching and students' approaches to learning. *Higher Education 37,* 57–70. https://doi.org/10.1023/A:1003548313194

Walsh, Z., Böhme, J., & Wamsler, C. (2020a). Towards a relational paradigm in sustainability research, practice and education. *Ambio.* https://doi.org/10.1007/s13280-020-01322-y.

Walsh, Z., Böhme, J., Lavell, B., & Wamsler, C. (2020b). Transformative education: Towards a relational, justice-oriented approach to sustainability. *International Journal of Sustainability in Higher Education.* https://doi.org/10.1108/IJSHE-05-2020-0176.

Wamsler, C. (2020). Education for sustainability: Fostering a more conscious society and transformation towards sustainability. *International Journal of Sustainability in Higher Education, 21*(1), 112–130, https://doi.org/10.1108/IJSHE-04-2019-0152

Wamsler, C., Brossmann, J., & Hendersson, H. et al. (2018). Mindfulness in sustainability science, practice, and teaching. *Sustain Science 13,* 143–162. https://doi.org/10.1007/s11625-017-0428-2

Wilber, K. (1999). *The Marriage of Sense and Soul: Integrating Science and Religion.* New York: Random House.

Zimmerman, J. & Virginia, C. (2009). *The Way of Council.* 2nd Edition, Las Vegas, Nev.: Bramble Books.

# 6

# LIFE: LISTEN. IDENTIFY. FEEL. EXCHANGE

## Bringing connection through emotions to students and leaders

*Sheila Sampaio*

### Introduction

About ten years ago, I participated in a group of studies on empathy and was introduced to the activity that completely transformed my perception on the topic. As a professor of MBA and postgraduate courses, and also as a teacher at the Sonata Leadership Academy – a school focused on adult education and self-knowledge that educates leaders through the Transformative Learning methodology (Sterling, 2011) – I was able to apply this activity to groups of young adults and business leaders from different sectors in Brazil. The classes in which the activity was applied were designed for professionals, managers and leaders of companies, whether in in-company training or in MBA and postgraduate business administration (for the most part) courses.

Most if not all of us have heard, at some point, that empathy is different from sympathy. It's interesting to note that in the last decade, the frequency of Internet searches for the word "empathy" has more than doubled, while searches for "sympathy" have fallen by approximately 30 percent.

We are most likely familiar, as well, with the concept that empathy is the ability to put yourself in someone else's shoes. I don't think this concept is wrong, just incomplete.

When trying to explain this skill – yes, because empathy is a skill and not a feeling and, in this way, it can be learned and developed – it's not enough to say, "Put yourself in someone else's shoes." Empathy means taking the perspective of the other and feeling an emotional bond with that person. Empathy has both emotional and cognitive components. The emotional component of empathy relates to *experiencing* another person's emotional responses; the cognitive component refers to the capacity of *understanding* others' emotions (Sevillano et al., 2007).

In his book *Empathy*, Roman Krznaric described empathy as the art of putting oneself in the other's shoes through imagination, understanding their feelings and perspectives and using that understanding to guide their own actions (Krznaric, 2015). In other words, there is no possibility of putting ourselves in the other's shoes if not through imagination, because the other's shoes are known only to the other. However, the goal of imagination is to connect with and understand

DOI: 10.4324/9781003229735-7

the other person's feelings and perspectives. This understanding then impacts our actions and the way we communicate, because the results of our empathy are only manifested through action.

Thus, any effort to understand empathy and develop the skill of empathy will be in vain if our focus is misplaced – that is, we connect with the situation in which the other person finds himself or herself and not with the emotion he or she feels. The dynamics presented in this chapter have proven to be an excellent tool to demonstrate this perspective to students – that empathy is a connection to emotion and feelings, not to a situation – in a practical way. I have been facilitating student development in empathy for more than 15 years, whether in educational institutions or companies, and I deeply believe that nothing can teach us more than experience and emotion.

## Why talk about empathy in a book of sustainability?

After hearing Isabel Rimanoczy talk about sustainable mindset, I understood that sustainability manifests itself in people and leaders acting for the common good – and soon realized that it would be very difficult for someone to wake up to a sustainable mindset without adopting an empathic posture. For individuals to care about a common good, they have to understand what that "good" means not only from their perspective, but also from the perspective of others. They have to care and value the different actors of their social group, who have different characteristics, interests, values and contexts.

The relationship between empathy and sustainability represents a key advance in understanding the drivers of human-environment relations. The lack of empathy for nature and for others limits one's motivation to conserve the environment and enhance sustainability (Brown, Adger, Devine-Wright, Anderies, Barr, Bousquet & Quinn, 2019).

The activity described in this chapter generates a connection with our own feelings, but also highlights the equal importance of attention and care for the feelings of others. It is an integral look at ourselves and the other. The objective of this activity is to develop in students an empathic attitude that will enable them to understand that there are other perspectives. To learn empathy, it is not enough to know the concept, one must experience it.

## Activity description – Life: Listen | identify | feel | exchange

I describe below the guidelines and steps followed to carry out the activity, which is divided into two stages: Preparation and Activity.

It should be noted that participants are invited to get in touch with their feelings and emotions, and to report them to other people. Thus, it is very important to create a safe environment in which they feel comfortable sharing such emotions.

Another important point is that the participants do not have prior knowledge of what the stages of the activity will be. These will be revealed as the activity develops.

### *First stage – Preparation*

*Step 1 – Questions*: First, I ask participants the following question: "Have you ever been in a situation where you felt that someone did not empathize with you?" or "Can you remember situations that would have been easier for you if the other person were more empathic?" They don't have to report the stories, just search their memories.

*Step 2 – The title*: Next, I ask them to think of a specific situation in which they felt a lack of empathy from other people, and to write on a piece of paper a title for this story (a word or a short sentence is enough).

*Step 3 – Forming pairs*: I now divide the participants into pairs (if required by the number of participants, we can have a trio).

## *Second stage – Activity*

*Step 4 – LIFE*: The participants are now instructed to tell their partners in their pairs the stories for which they wrote the title on the piece of paper, seeking especially to report their emotions and feelings. When the participants are listening to the stories of their partners, they should adopt a posture according to the following guidelines:

**L – listening**: Listening to their partners with genuine interest, with authentic listening.

**I – identifying emotions**: Trying to identify which emotions their partners express while telling the story.

**F – feeling and taking the other's perspective**: Trying to remember for each emotion reported by their partners, moments when they felt such emotions themselves.

**E – exchanging presumption for validation**: Before assuming what their partners were feeling, trying to validate if they have correctly understood their emotions. Asking questions like "Did you feel angry at this point?" or "And did that make you feel afraid?" will help to validate feelings and emotions.

*Step 5 – Exchanging titles*: As soon as both partners in the pairings have told their stories, I ask the pairs to swap titles. That is, the participants of the duo must hand the paper with the title of their stories to the other person.

*Step 6 – Changing pairs*: The participants are now instructed to change pairs. Each must form a new pair with another person in the group.

*Step 7 – Telling the story as if it were your own*: Finally, the participants must retell – to their new partners – the story they heard previously, as if it were their own. Here it is very important that they tell the story in the first person, as if they had lived the story they heard. The focus, again, should be on emotions and what they felt when it happened, especially with regard to the lack of empathy. Whoever is listening should, again, adopt the LIFE posture, described in step 4 of the activity.

*Step 8 – Group analysis*: At the end of the activity, everyone is invited to form a circle and share their experiences with the activity. This is the richest moment.

## *Illustration of the activity*

Anne and Max formed a pair. Anne told the story with the title "Vacation" to Max, and Max told the story with the title "Homework" to Anne. As soon as they finished telling each other their stories, Anne and Max exchanged their story titles.

Then Anne formed a new pair with Peter, and told him the story entitled "Homework" that she had received from Max as if it were her own story. She then listened to the story Peter had to tell her.

In the same way, Max formed a new pair with Paul, and told him, as if it was his story, the story he had received from Anne entitled "Vacation." He then listened to the story that Paul had to tell him.

## Activity analysis

As described above, in the preparatory stage of the activity (steps 1, 2 and 3), the participants had to remember a moment when they felt that someone did not empathize with them, choose a title for this story and get into pairs.

Note that even if they answer in the affirmative to the step 1 question – "Have you ever been in a situation in which you felt that someone did not empathize with you?" – many participants have difficulty with step 2: specifically remembering a moment when they suffered the impact of a lack of empathy, and giving the story of this moment a title.

This demonstrates that, in general, students remember the emotion of feeling the lack of empathy; however, they have difficulty in classifying situations as such. Alternatively, they might find it difficult to choose a situation that they feel comfortable sharing. In any case, this is the first moment of the activity in which they come into contact with their emotions. Note, also, that at this moment, some participants quickly remember and choose their stories, while others need more time. It is up to the advisor to lead the group so that the participants do not feel frustrated.

Step 3 usually takes place smoothly. Here, to make the experience more interesting, I suggest that the participants get into pairs with people with whom they are less familiar – but that it is only a suggestion.

In step 4, when students are instructed to share their stories focusing on their emotions and feelings, they generally report a mixture of excitement and fear, which are natural reactions in activities that involve sharing something so personal. In addition, for this profile of students, which include professionals, leaders and managers, the pattern of communication and reporting of facts is more pragmatic, technical and objective – that is, much less related or interested in feelings and emotions. The simple fact of putting themselves in a situation in which they need to report feelings and emotions takes them out of their comfort zone and puts them in a position of vulnerability. As Brené Brown says, shame, or a sense of personal unworthiness, impedes connection (Brown, 2010). This means that overcoming any possible shame and allowing themselves to be vulnerable by sharing feelings with a colleague, is the key to making a connection.

In terms of listening to the stories in step 4, I need to explain how these stories should be heard through LIFE (Listening, Identifying emotions, Feeling and taking another's perspective, and Exchanging presumption for validation). This explanation takes time as most people are not used to providing this level of attention when someone is speaking to them.

Most groups react at the end of step 4 as if it were the end of the activity. The expectation is that the activity will be limited to the exposure of their stories and the exchange of experience through empathetic listening.

It is true that the activity could end in step 4 with the possibility of a rich exchange of experiences: how they felt sharing their stories; how they felt expressing themselves with a focus on their emotions; the experience of listening carefully to their colleagues; how empathic listening was perceived; etc.

However, the big moment is yet to come.

Step 5, swapping the titles of the stories, usually comes as a surprise: "But isn't the activity over?" or, "Is there still more?" they say. These types of questions continue when I give the instructions to switch pairs. In step 6, I always welcome with a wide smile the curious and suspicious looks that are thrown at me. The activity seemed simple and under control – a control they had just lost. Now they must form a new pair, and have with them the title of a story that is not theirs.

As I move on to step 7 and explain that they must now tell their colleagues the story they heard as if it were their own, the following response from a sales professional is what I typically hear: *"But if I had known that I would have to tell the story I was listening to, I would have paid more attention!"*

Even with all the guidelines on LIFE (Listen. Identify. Feel. Exchange), I have heard observations like this one more than a few times in the years of conducting this activity. I have observed that attentive listening is not part of the daily repertoire of most professionals, leaders and managers. A conscious and purposeful effort is always necessary to fully pay attention to the other person – and that becomes very clear in this stage of the activity.

There are also situations that we can consider fun or funny and some that generate doubts. For example, suppose a member of the pair is a man who heard a story from a woman in which she did not notice empathy from others while pregnant. How will he tell this story? The same is true in the case of a woman who must tell the story of a man who did not notice any empathy in the way his girlfriend broke up with him. Should she tell the story about her girlfriend or move on to a boyfriend?

The truth is that this type of situation brings great opportunities for learning. And the answer to the dilemmas described above is for participants to assume the story as if it were theirs, and remember to focus on the emotions and feelings when reporting it, without changing anything. The idea here is that if we think that we can empathize only with situations that we have lived or that we have the possibility of living, we will hardly be able to exercise real empathy for someone else. We will never be able to live the same experiences of all the people with whom we relate; and yet, by focusing on their feelings, we are fully able to connect through empathy.

After the time set for step 7 has elapsed, everyone is invited to form a circle to start the last step: step 8. I recommend that you do not rush and that you allow the group enough time for these reflections.

You can provoke the group with questions such as, "How did it feel to share your story?," "How did it feel to hear the other's story?" and "Do you agree it was a matter of lack of empathy?" These types of questions will elicit many reflections, and the richness of this moment lies in the group's own diversity.

Here are some examples of what I have heard from participants:

"*When I heard the story my colleague told me, at first I thought it was silly and unimportant. But when I had to take the story as mine and I needed to connect with the emotions that she reported she had felt, I started to understand that, yes, it was an important issue and a lack of empathy*," a manager once reported to the group.

We are in the habit of judging the emotions and actions of others based on our own experiences and references, and this takes us away entirely from the potential for empathy. Likewise, it keeps us from developing connections when we do not have the internal resources necessary to consider important what is different from us and what we are not always able to understand.

"*I think it was the first time that, in fact, I listened to someone carefully. And I thought I was always paying attention*," wrote a graduate student.

The possibility of being in a safe environment and with an exclusive focus on caring for others, without distractions, allows students to live the connection through listening, and raises their awareness of whether, in their daily lives, they actually practice this attention. And when we learn to pay attention to emotions and feelings, and not just the information transmitted, we are able to optimize communication and generate better results. Often, what we do on a daily basis is not listening, it is just allowing other people to speak without interrupting them, and that is not paying attention.

"*I realized that I have a hard time talking about emotions. When telling my story, my colleague was asking me, 'And what did you feel?' or 'And how did you feel after that?' I think I avoid this kind of conversation*," reported an IT executive.

Organizations want people to work as a team and communicate assertively, but they do not create environments that stimulate interest or encourage the sharing of emotions, which is the basis for human connections. If we are unable to value and express our feelings, how can we recognize and be interested in the feelings of others?

"*I thought it would be more of a group dynamic to generate networking, but it really wasn't. I realized how difficult it is to put yourself in the other's shoes. Only those who went through the situations know, in fact, what they felt. The closest I got was trying to remember when I felt this way, but I know it's not the same*," said a project manager leader.

Empathy has a limit. It is that simple. And recognizing that we will not always be able to understand what the other is feeling frees us from the requirement to have answers for everything. Empathy is cited as one of the most important soft skills for professionals in all areas, but we need to know what our limit is and be aware that we will not always have the internal resources and the ability to connect empathetically. And that opens up the possibility that others will not be able to empathize with us either.

## Students' reflections

As this is a very practical activity and there is an incentive to speak and listen for group sharing, students are invited to freely write about their personal experience during the activity. This is not mandatory. All the reflections cited here are from Brazilian students, written in Portuguese, and were translated with an effort to stay faithful to their emotions. Here are some of their reflections:

> *I relived each moment by retelling my story to my colleague, and that only reinforced my conviction of how important it is to show feelings. When you are true, you transmit the intentions that come from your heart, only someone very insensitive can repudiate them. And, in the face of all the emotion in my story, I devoted total attention to listening to my partner's story, assuming it as if it were mine, since it also overflowed with feelings. Evidently, each human being feels and reacts differently, which is the reason for the challenges of the activity, as practicing empathy is not an easy task. It can and it must be practiced.*
>
> *The synergy of the participants was noticeable, the desire that everything we were learning could be replicated in our jobs, reverberating throughout the classroom.*
>
> *The activity brought adrenaline to the surface and the opportunities to raise our level, as leaders, simply flowed. The fact that we have to take someone else's place, by itself does not make us empathetic. I think that the activity brought much more than that, it gave us the chance to be better people and leaders.*
>
> *Student 1, Sonata Brasil*

> *We are often listening to people talk, just listening ... After that activity, I realized how much we have to be available to people, so that we can really feel what the other person wants to tell us, even if we don't say anything, but the person feels that we are actually present. Our lives are made up of relationships and we need to be aware of this. At that moment, I felt how important vulnerability is, to feel that the other person is being genuine in his story. That we all go through difficult times. It takes sensitivity, active listening and, above all, presence and time.*
>
> *Student 2, Sonata Brasil*

> *I always had good expectations, I knew something would come that would move us. Telling a story that is not yours, and trying to feel what the person felt when telling the story is not easy. At this moment, we saw the difficulty of putting ourselves in the shoes of others in some situations. It was quite challenging. Certainly everyone wanted to tell the story with the maximum similarity to the original. I realized the challenge of really understanding that everyone has a way of feeling things. What can be little for one, can be a lot for the other, at this moment we must know how to respect both.*
>
> *Student 3, Sonata Brasil*

> *I was very curious to understand what it would feel like to tell someone else's story. It was 'strange' to have to incorporate a feeling about something that hadn't happened to me. People's behavior in some situations can often be explained by things that have already happened in their lives, and that are not necessarily related to the current situation itself. I think the activity was important to show people who look only superficially at others that their 'modus operandi' often comes from something that happened a long time ago.*
>
> *Student 4, Sonata Brasil*

> *So, I really expected to be able to cope with the other's emotion when telling the story. I was very nervous and hoping to finish soon. I was afraid I was telling it wrong. Taking on someone's story was very difficult, almost impossible. It made me think about how much we judge without knowing the other person's feelings. From this activity I intend to think much more before acting, to consider all possibilities when charging someone for something. Open mind for explanations.*
>
> *Student 5, Sonata Brasil*

I could report numerous excerpts from students' assessments of the activity, the vast majority of which refer to becoming aware of expectations about their real experiences that emerged from the activity, becoming aware of the importance of listening, or realizing how empathy is something challenging to develop in relationships.

While the students only *listened* to the stories and feelings reported by their peers, there was room for judgment and comparison. But when they were asked to connect to the stories they heard through their own feelings, then there was room for real empathic understanding.

The main lesson we can learn from this activity is that the empathic process stems from the connection through emotion, not the situation. If we try to understand what other people are feeling from the situation in which they find themselves, we will probably judge their feelings and emotions based on how we would react or feel ourselves. However, each person feels and reacts differently to situations. Thus, we are able to understand their reactions if our attempt at understanding is based on feelings about others that we have experienced at some point (even if the situation was different).

## Empathy for a sustainable mindset

The presented activity allows students to experience the real dynamic of self-knowledge, connecting with themselves while paying attention and caring for the other's perspective. If certain people are unable to connect with others as human beings, how can we expect them to connect with their communities, or with a common good, or with our planet?

Developing an empathic attitude also means understanding our responsibilities as professionals, managers or leaders, and the impact that our way of relating and communicating can have on our teams and the organizations of which we are a part. And what is the responsibility of a leader if it is not to have a sustainable mindset? In addition, through the activity, we can have self-awareness including the awareness that we, perhaps, are not in the habit of actually listening to people. We learn to recognize our attention gaps.

Empathy is likely to be important in three different dimensions of sustainability action. First, empathy is required for individuals to understand the consequences of the impacts of environmental change on the natural world and on other populations. Second, for empathy to contribute to sustainability, it requires conceptions of place, community and identity beyond those based directly on kinship and immediate locality. Third, the extent of empathy determines the ability and likelihood of coordinated collective action for pro-environmental decisions. Thus, the presence of empathy generates the moral impetus for action (Brown et al., 2019).

Being able to observe, listen, and be interested in perspectives other than our own allows us to move from an individual focus to a focus on the collective – and, thus, to put ourselves in the proactive stance of thinking about solutions for more sustainable actions of all kinds: a species of animal facing extinction, the polluted water of a river, a community in a situation of vulnerability, the destination of residues, access to drinking water or better working conditions. The possibilities are endless.

> *We don't necessarily need to go very far or accomplish great things, because we relate to people all the time, and each interaction allows us to exercise sustainability.*
>
> *Student 6, Sonata Brasil*

Having a sustainable mindset requires the development of awareness. Although empathetic people create corporations, many businesses operate in a non-empathetic manner (Patnaik, 2009). Companies are important actors in our social system, and their management model – based on what is given focus and importance – has the potential to significantly impact people's behavior (employees, suppliers, customers, society). Workplace management and workers can all play a role in promoting empathy and environmental policy. At the management level, empathy might be promoted by policies that encourage employee collaboration, teamwork and mutual problem-solving that foster perspective taking and empathic concern (Di Fabio & Kenny, 2018). Having professionals capable of paying attention and really listening to the wishes and needs of the actors around them can be a great differential in the development of a more sustainable society.

## References

Brown, B. (2010). *Brené Brown: The Power of Vulnerability*. TED [Video file]. Retrieved from https://www.ted.com/talks/brene_brown_the_power_of_vulnerability?language=pt

Brown, K., Adger, W. N., Devine-Wright, P., Anderies, J. M., Barr, S., Bousquet, F., ... & Quinn, T. (2019). Empathy, place and identity interactions for sustainability. *Global Environmental Change, 56*, 11–17.
Di Fabio, A., & Kenny, M. E. (2021). Connectedness to nature, personality traits and empathy from a sustainability perspective. *Current Psychology, 40*(3), 1095–1106. doi:10.1007/s12144-018-0031-4
Krznaric, R. (2015). *Empathy: Why It Matters, and How to Get It.* TarcherPerigee.
Patnaik, D. (2009). *Wired to Care: How Companies Prosper When They Create Widespread Empathy.* Ft Press.
Sevillano, V., Aragonés, J.I., & Schultz, P.W. (2007). Perspective taking, environmental concern, and the moderating role of dispositional empathy. *Environment and Behavior, 39*(5), 685–705.
Sterling, S. (2011). Transformative learning and sustainability: Sketching the conceptual ground. *Learning and Teaching in Higher Education, 5*(11), 17–33.

# 7

# A SEMESTER WITH A TREE

## Meaning making and global learning contexts

*Michael Lees*

### Introduction

Undergraduate students encounter individual, local and global pressures that negatively affect awareness for self, other and the environment. Figure 7.1 illustrates student concerns with these pressures[1]. Chickering, Dalton and Stamm (2006) maintained that students currently experience lack of stress relief, burnout, rising costs of living, apathetic productivity and loss of time in life for personal reflection and social connections. Safranski (2005) asserted contemporary globalization pressures include terrorism, environmental degradation and industrial and cultural exploitation, as overarching elements influencing individual experience. At the time of this writing in 2020, the Covid-19 virus has infected millions and killed over one million people. Students are contending with a major local and global paradigm shift as college learning has temporarily moved online to counter a pandemic's effects. Juggling stressors involving their personal, social and global lives has moved to the foreground. The opportunity to digest personal experience, with continued learning outcomes, becomes an important part of the learning process.

My intention in designing a Tree Journal exercise supplements the development of student spiritual, ecoliterate and sustainability mindsets as it concerns individual, local and global learning contexts. The Tree Journal (Appendix 1) requires that students befriend a tree that they will spend the semester having conversations with. The journal consists of question prompts aligned within a spiritual pedagogical framework that required the students to ask the tree what it thinks about the questions found in the journal guidelines. They are tasked with recording the tree's responses. Students recorded these responses, reflected on the responses, and intermittently presented their tree-friend's views on life in class throughout a college semester.

### Globalization: Individual and global learning pressures in undergraduate student life

Individual, local and global fragmentation creates confusion, anxiety and fear in the life-experience of students. Fragmentation in the lives of students is addressed by Student 2: Montclair State University's Tree Journal entry:

DOI: 10.4324/9781003229735-8

**FIGURE 7.1** Student 1, Montclair State University: Tree Journal. Photograph by Michael Lees

> *Sammy's (Tree Name) attitude towards life is pretty complex. A lot of events that happened during his lifetime have led him to have negative thoughts towards humans, but an open-mind to the way they think. He has a lot of life-changing ideas that he wants to tell people … he is sick and tired of being a tree because people aren't appreciative of him and he has no voice. Sammy's philosophy on life is to be who you want to be, not to be what the world wants you to be.*
>
> *(Student 2, Montclair State University)*

Hugonnier (2007) outlined five major challenges that globalization poses and can subsequently lead to individual fragmentation and feelings of being overwhelmed:

- Global economic interdependence now leads to greater international competition and distribution of labor
- An increase in economic exchanges has led to the need to develop awareness for cultural sensitivity, languages, religions, races and social capital
- Social interaction and income levels have changed drastically, negatively effecting social unity in local, national and international contexts
- Environmental degradation in terms of ecology, poverty, drugs, arms and human relationships stymies the recognition of a participatory global citizenship
- The recent influx of attention to the internationalizing of higher education risks students being exposed to opportunity inequality, poor education or non-relevant learning.

**FIGURE 7.2** Student 1, Montclair State University: Tree Journal Entry. Photograph by Michael Lees

Undergraduate students currently find themselves immersed in these challenges and are generally left to figure out how to work their individual lives into a global context on their own. This is reflected in the Tree Journal Entry of Student 3, Montclair State University:

> *Late nights bring fights, crises, and strife – for the people who are truly awake. Alisoun (Tree Name) finds there is no person more like itself than the teenager up until two hours before dawn, mindlessly observing – not consuming – endless streams of information. Yes, they are very similar. Alisoun likes to think that it retains more than it lets pass by, but surmises that its memory has been shot since it lost its branches.*
>
> *(Student 3, Montclair State University)*

Bringing awareness to these underlying stressors acting on a student's psyche requires attention be paid to global consciousness and global learning (Figure 7.2).

## Global consciousness and global learning

Contemporary undergraduate students represent a body of emerging adults that encounter large amounts of technology, media and information in their studies and daily lives. Information of this magnitude leads to a personal feeling of disconnection, helplessness, and lack of power as individual lifeworld views grow narrower (Safranski, 2005). As Student 4, Montclair State University's Journal reflects:

> *Ollie (Tree Name) doesn't think about all that he cannot do. "What can I, a mere puny tree that is cemented to the ground possibly do?" Oliver once asked me. With a small smile on my face, I stood up from the dirt and held onto one of his small thin branches. I told him that he gives life to this world and its inhabitants. His magnificent leaves give us air and his long branches provide homes for smaller creatures. I told him that with his mighty roots that burrow the ground he makes our ground and soil smore stable.*
>
> *(Student 4, Montclair State University)*

Mansilla and Gardner (2007) argued that young adults need to better understand the multi-faceted information and patterns that influence individual, local and global contexts. A deeper understanding includes stepping beyond basic educational knowledge acquisition by connecting students with direct experiential aspects of global learning (Mansilla & Gardner, 2007). This means teaching students how to read the world, and world's interdependent systems, using an ecoliterate approach (Lees, 2017). The development of ecoliterate conscious awareness affords the opportunity to establish connections between personal life actions affecting individual and global relationships.

Creating a space for students to ask questions externally, questions they might never ask outside themselves at all, supports a big-picture, or ecoliterate lens. Ecoliterate means being able to read the world, gain knowledge, and do something with that knowledge among all the personal, local and global systems at play. Student 5, Montclair State University's Journal entry illustrates this ecoliterate lens:

> *There are so many things happening in our everyday life that we want to share our experiences to someone. I think we need at least one person to tell our secrets in this busy world. Though, I would not like to disturb anyone because I know everyone is busy with their personal life. However, a best friend is an important person that you can trust and tell anything even your biggest secrets. I believe there is only one person in my life and that is my imaginary tree friend "Krishna." Therefore, I could simply share my feelings with him and tell him how I spent my day. The hidden life of the trees is they can communicate, feel and have emotions with each other and human beings. Similarly, Krishna would like to listen to me and keep my secrets.*
>
> *In my view, Krishna always has a positive attitude towards life. The crucial reason is that he provides us shelter, shade, and nutrition to protect individuals from the storm, rain, or other environmental danger. In addition, he fulfills all human needs but he never asks for anything. He will always be there standing to fight for human beings. Moreover, trees are the key of health and happiness. Because Krishna takes all my worries, fears, tension, and anxiety when I am stressed out by telling him all of my world problems. In this way, he provides me support, confidence, and love during my hard times. Therefore, having a tree friend is both physically and emotionally helpful for me.*
>
> *(Student 5, Montclair State University)*

Contemporary forces of globalism are calling for higher education to address how students can be provided with the right set of skills to address the development of their individual and global awareness. Mansilla and Gardner (2007) promoted four points that educators can use to focus student learning on the development of global consciousness:

- Economic integration examines the opportunities and costs relative to societies, cultures and individuals involved in producing capital

- Environmental stewardship addresses global ecologies, health and sustainability
- Cultural encounters situated to examine the relationships of cultural exchange in globalization
- Governance in understanding national and international forms of governments, global rights and global citizenship.

The use of these four focus points begins to provide a student with the capacity to see and understand the larger picture of globalism in action. Although these four points do begin to build student worldviews, they do not address the individual in relationship to globalism in action. Global consciousness and current global education approaches address the outer development of student awareness in a knowledge-based and objective style that is grounded in empirical methodologies. The use of current approaches to learning in undergraduate education centers on knowledge in theoretical, empirical and segregated disciplinary lines of transmission (Chickering et al., 2006). The focus on the transmission of knowledge and the student as consumer has left students feeling bereft of a well-rounded education that prepares them for their individual and global life experiences (Chickering et al., 2006). Palmer (1993) argued that if global collaboration is to become a pluralistic reality, then truth seeking between objectivism and subjectivism in education needs to take place. A gap exists in which whole student maturation needs to find further address in balancing objective and subjective learning experiences for an ecoliteracy-based global context. The next section introduces spiritual pedagogy as a means to balance objectivism and subjectivism in student approaches to learning.

## Considering spiritual pedagogy

If higher education provides students with a holistic global learning environment an opportunity arises to ground their education in intellectual, psychological, social and environmental development. Astin, Astin and Lindholm (2011) asked "If students' lack self-understanding – the capacity to see themselves clearly and honestly and to understand why they feel and act as they do – then how can we expect them to become responsible parents, professionals, and citizens" (p.2). Chickering et al. (2006) contended that if higher education continues to only emphasize transmission of theories, empirical learning and objectivism, without attention to wisdom, compassion and integrity, students will not obtain the necessary skills to develop holistically. Wisdom, compassion and integrity include justice, ethics, values, morality, virtues and character thus offering students the ability to ask why they are learning what they are learning (Chickering et al., 2006). The importance of students asking, "Why learn this stuff?," embeds itself in the academic journey of knowledge acquisition via meaning making and purpose.

Spiritual pedagogy (SP) provides the ability for educators to engage students in a process of integrating knowledge with meaning making and purpose. SP offers educators the means to enhance the subjective development of students' lives by bridging individual subjective experiences with objective knowledge. Spiritual pedagogy involves exploring meaning making and seeks to connect the affective experiences with the logic and reason students are gaining (Astin et al., 2011; Chickering et al., 2006; Nash & Murray, 2010; Palmer, 1993). Subjective learning experiences involve felt responses to personal, local and global issues. Objective learning involves knowledge for the sake of knowledge. SP helps build a bridge between subjective experience and objective knowledge with the overarching intention to see the pragmatic fruition of wisdom as learning outcomes. This will not happen without the opportunity for students to reflect, digest and contemplate learning contextually.

Essentially, how does a student take what is learned in the classroom out the door with them at the end of each class and semester? How do they find meaning and practical application in what they have learned? Seeking personal understanding and meaning is reflected in the Tree Journal entry of Student 6, Montclair State University:

> *Perched in front of my college hall, Oliver watches solemnly. How he got there he knows not. He only remembers the now, and what happens in his long, day to day life. He knows that he once spawned from a seed in the ground, but his past history is something he can no longer remember. Hundreds of students pass him hourly, yet none of them stop to look at him. None of them stop to speak to him nor to listen to him. He's lonely, like me, and only waits for someone to notice him. He longs for interactions, for someone to understand him. No birds found him hospitable to make him their home and he bears no fruit. Other humans decorated him with strings and bright heavy lights once he became bare and shed all of his captivating leaves. This makes him feel not good enough. Like he needed to change for humans to fully admire him. But I didn't notice him for his light. No, I chose him for his unique shape; for the way it looks like he's desperately trying to reach the sky and grow beyond this planet. I wish I were that courageous. I wish I had the determination necessary to try and transform this silly fantasy into a majestic reality. However, like the rest of us, he stays put; only growing deeper into the earth and spreading further away from where he wishes to peak.*
>
> *(Student 6, Montclair State University)*

Meaning making is a continuous process for students as they progress through academic course loads and college life. Student engagement of meaning making will cycle through various choices based on the encounters present in their personal, social and academic life. Semester after semester students have approached me about choices they are making concerning switching majors, keeping a major, outcomes concerning a major, all of which are relative to personal, local and global concerns. Internal personal questions and external environmental reactions suggest that causality plays a role in making choices. The search for meaning as it concerns outcomes based on their choices allows students to realize they are part of systems. Systems where personal activities and outcomes have a mutual effect on individual, local and global levels. The individual process of working with choice, trial, error, success and failure, exists as a personal journey in which learning how to relate to self, other and the world is taking place.

Undergraduate student learning involves processes of self-discovery and transformation. This requires helping students open up to who, what, when, where and why they are learning what they are learning relative to their personal, local and global experiences with the world. How do students apply what they have learned becomes an important question for educators and the halls of academia to address. Students that take the time to reflect on how to connect their objective knowledge acquired in the classroom with inner subjective responses to learning begin to find ways of creating balance amidst their busy lives. The search for undergraduate students to find balance in the fast pace of individual, local and global learning experiences requires equanimity. Equanimity is marked by achieving a state of calm, simplicity and remembering to breathe. A spiritually pedagogical approach to learning aids students in being able to work with the hardships, difficulties and challenges that are part their academic career. Finding time for students to foster equanimity is what led me to design the Tree Journal.

Ultimately, students are seeking to achieve successful life and academic outcomes as they progress through their four years of undergraduate studies. Students choosing to participate in a college education enter a world that can appear to be overwhelming and at the same time narrow and isolating. Chickering et al. (2006) stated, "Higher education needs to find a way to speak with conviction and compassion to today's aimless, yet desperately searching, Gen-xers, Millenials, and Quarterlifers" (p. 39). Glazer (in Chickering et al., 2006) adamantly argued that students in today's world are participating in a society and education that includes a limited and materialistically saturated worldview. Glazer claimed this worldview teaches students to horde and protect individual gain versus sensitize and collaborate for global betterment. Students, at the undergraduate level, are emerging adults trying to figure out how to be in the world. Empowering their search, encouraging their questions and serving as a guidepost provides opportunities for students to thrive. Student 7, Montclair State University's Journal Entry reflected on finding identity within a world that is in constant flux, change, and turmoil with an illustrated sense of inner identification, strength, and purpose (Figure 7.3).

**FIGURE 7.3** Student 7, Montclair State University: Tree Journal. Photograph by Michael Lees

The life experience of the year 2020CE directly exemplifies the need for a pedagogy that provides a ground for connecting personal, local and global learning. The establishment of space that allows for analytical and introspective critical inquiry provides students with the ability to learn authentically while expanding their worldview. The use of a journaling exercise, throughout the duration of a course semester, provides the ability to work with a spiritual pedagogical approach to learning and the development of a sustainability mindset in the classroom.

## The Tree Journal

Journaling provides students with a means to explore their thoughts and ideas with critical inquiry, creativity and reflection relative to course material. Students have the open opportunity to explore teachings and foster a sustainable connection to the subject material being studied. Spiritual pedagogy (SP) is about meaning making, purpose motivation, connection and the establishment of a sustainability mindset. The sustainability mindset serves to integrate what is learned objectively, subjectively, inside, and outside, a college classroom. A sustainability mindset's core competencies include, "systems thinking, self-awareness and spirituality within the dimensional contexts of being (values), thinking (knowledge) and doing (competency)." (Kassel, Rimanoczy & Mitchell, 2016). Bolstering the sustainability mindset core competencies are supported in the establishment of a meta-spiritual pedagogical approach in which objective and subjective learning have an opportunity to connect in the mind of a learner. This directly relates to their overall experience in a college environment and classroom. So what does meta-spiritual mean?

Welch and Koth (2013) outlined a metatheory of spirituality that defines the connective moments of objective (cognitive) and subjective (affective) experiences as taking place in relational spaces of thinking, doing and learning. Welch and Koth (2013) define space as a place where students can explore, delve and evolve versus traditional linear approaches to learning and knowledge acquisition. The creation of an open and exploratory space allows spiritual development to happen in phases rather than stages so that integration occurs with a dynamic fluidity absent of rigid structures (Welch & Koth, 2013). I created Figure 7.4 to illustrate the six spaces of a metatheory of spirituality that students encounter during their college years:

Note: Figure 7.4 represents the relational spaces that students encounter during their college experiences. Adapted from "A Metatheory of Spiritual Formation Through Service-Learning in Higher Education," by M. Welch, and K. Koth, 2013, *Journal of College Student Development*, *54*(6), pp. 616–618.

Attention to these relational spaces by members of higher education communities provides students with the ability to cultivate their spiritual development and provide them with authentic learning experiences (Welch & Koth, 2013). The authentic learning experiences connect subject knowledge, affective awareness, and potential moments for transformation for students with individual, local and global learning.

Journaling provides moments for students to reflect on their relationships to what they are learning on personal, local, and global levels. Having this opportunity to reflect fosters a sustainability mindset by providing a safe ground for students to be comfortable, express themselves personally and confidentially, and situate what they are learning as something relatable to their own life experience. Student 8, Montclair State University's Tree Journal reflects this when the student stated:

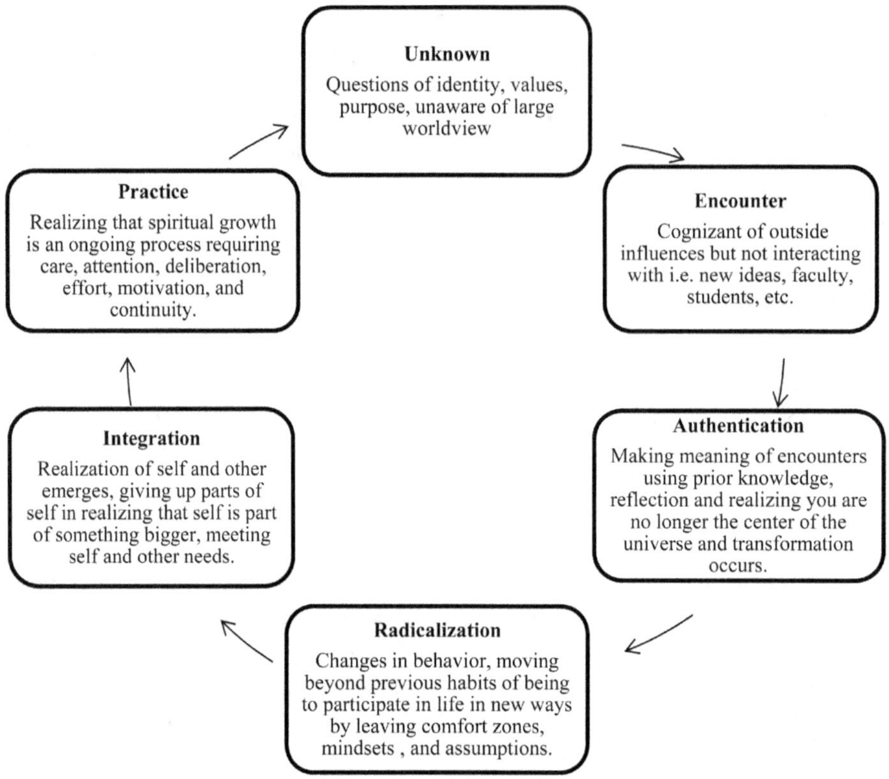


**FIGURE 7.4** Six Relational Spaces of a Metatheory of Spirituality

> *My visitations towards seeing Christopher weren't as consistent as I would personally want them to be. It was more so when I needed to go somewhere to find my peace of mind or even so just to have an escape. I enjoyed his company as he made me feel like he was awakening my third eye. There wasn't a moment where we had an awkward silence, but instead we didn't have enough time to say everything we each had to put on the table.*
>
> *Christopher loved windy days as his leaves and branches would begin to shake sounding like the beat to a catchy pop song. His energy was so vibrant, everything he would do or say would be different each time. To speaking about the findings beyond this galaxy or being active. Even him as a tree, he still found ways to keep motivated. However we always mostly kept focus on one particular topic: How are government systems so unhealthy? He, who we depend most on, humans needing trees for oxygen, we have forgotten that. As we are making the trees disappear.*
>
> *(Student 8, Montclair State University)*

The end result of journaling in a classroom involves the promotion of establishing systemic connections and a sustainability mindset to personal, social, global and environmental constructs. This kind of awareness is ecoliteracy. The four aspects of personal, social, global and environmental constructs via an ecoliterate lens encapsulate the essence of the purpose and motivationally driven theoretical foundation of a spiritual pedagogical approach to learning. Student 9, Montclair State University illustrated ecoliterate awareness in their journal:

> *I found a tree on my property and I introduced myself long ago. The tree is named Olaf and he is a pine tree. When I was younger I planted many trees around my house. Olaf was one of ten trees that grew and reached the top of my bedroom window, for me to introduce myself. My tree friend makes sure I have shade in the morning, so the sun doesn't blind me when I am still asleep. Olaf keeps birds in his branches so other life can grow. During sunset when I do most of my work he helps me study by keeping me company and providing shade so the sun doesn't blind me through the window. He produces a lot of pinecones as well for the animals to feed off from. Olaf likes to give back to the world and he tells me he is here for that reason. Olaf is now 16 years old and he is still prospering.*
>
> *(Student 9, Montclair State University)*

Using a tree provides a neutral and safe space in which the learning taking place is student centered. The use of a tree as friend, listener, talker and companion opens the space and invites students to share in an authentic way of looking at themselves in relationship to the world.

The assignment requires students to keep this journal via assigned due dates throughout a college semester. The assignment questions provided throughout the journal raise numerous issues that students are facing in their own emerging adult lives. The opportunity to engage in these questions with a "tree-friend" provides students a sounding board that reflects their own thoughts, ideas, and projections about themselves and life, with someone that is "there to listen." The listening tree friend offers an unbiased, non-prejudiced medium that affords students the ability to explore some of life's deeper question without reservation or fear. These questions provide the ability for students to outwardly reflect on internal questions of spiritual development. Curiosity plays a role in how learners decide to keep or reject what is being learned. Trusting processes of learning consists of a unique and tenable relationship between self, other, and world. This inner-questioning is expressed in Student 10, Montclair State University's Journal entry:

> *Alisoun (Tree Name) trusts the earth, but not the sky, because the sky is what sent the hurricane that ultimately killed it and its mother. Although Alisoun is not entirely dead, it likes to think of itself as a ghost lingering – this is a fantasy that also subdues fear when a new storm is swept into its sky. Lacking a top makes it both more stable and more hidden.*
>
> *I wondered if the sky beyond the troposphere was redeemable in Alisoun's eyes. It said it couldn't' see as far as that but assumed it must be one of the most peaceful places to be, perhaps too peaceful. It doesn't want to lose touch with the hard ground, the network that reminds it that there is a world outside of itself. In short, it would rather live in the world than in its mind even if it can't see as much of it as could be seen from that greater distance.*
>
> *(Student 10, Montclair State University)*

The Tree Journal is student centered in how they choose to construct their responses. In order to accommodate student willingness in engaging in this project, students are given the choice to use a number of different mediums that include: writing, drawing, sketching, prose,

FIGURE 7.5 Student 11, Montclair State University: Tree Journal. Photograph by Michael Lees

poetry, painting, music, dance, video, pictures or a presentation (Figure 7.4). Additionally, the lead instructor for the course should also participate in this exercise with the students as well. This provides a lead-by-example model and the opportunity to further in-class discussions and coursework. The material in the Tree Journal offers the lead instructor the ability to use the questions and literature for in-class discussion, course construction and potential exercises as well (Figure 7.5).

## Conclusion

I have used the Tree Journal exercise in my classes for more than a decade. The student entries found within this chapter represent learning artifacts spanning that time. I have also participated in this exercise with students every semester. I now have a lot of tree friends on campus that I still spend time with daily. I share in my drawings, expressions and representations of the trees throughout the semester and respond to the questions in kind. As an educator I have found that participating in this myself, along with my students, continuously keeps me up to date with how students are doing and what students are dealing with in their personal, local and global lives. It has been a guide for me as an educator to stay with the pulse of culture found among the student population. Thus guiding my own curriculum design to meet with changing times and content construction. The sustainability mindset is also fostered in the educator with this exercise. Malleability, flexibility, change and systemic address are found in how I conduct and relate to course material, meaning making and the students in the classroom.

Meeting the demands of today's global era, while simultaneously addressing the pressures that these demands are creating in the lives of college students and educators, requires that higher education direct more attention to balancing student cognitive knowledge acquisition with affective social, emotional, spiritual and ecoliterate development. The 21st century global demands are affecting incoming college freshman in how they view their own personal sense of purpose and direction amidst highly driven individual, local and global competition. Since students perceive their academic engagements as a journey, a window-of-opportunity for higher education to institute spiritual pedagogy exists as a way to integrate student spiritual development with information processing. Student 12, Montclair State University said this well in their Tree Journal entry when reflecting on the overall experience of spending a semester with a tree:

> *It was enlightening and fascinating to ponder more about the world and dig at the knowledge that has been attained from being in it. It has made me realize how monumental one little thing like a tree can be and that you must understand your importance and value in life and to love yourself and love others while they're still here. It would definitely make a sound if we fell together in the world as it has touched others and made memories with plenty.*
>
> *(Student 12, Montclair State University)*

Providing an environment that is open, flexible and hospitable with an exercise like the Tree Journal, creates the opportunity to foster student ecoliteracy and a sustainability mindset for the journey young academics are undertaking in their higher education career. It provides students the ability to address issues in their own lives in real time. To take what they are learning and find applicability to life's situations as they arise. From Student 13, Montclair State University's Tree Journal entry:

> *Spending a semester with my tree friend was unfortunately cut short due to the Coronavirus. Students were evacuated from their dorms, sent home, and all people were to be quarantined. My tree friend taught me to have a more open and willing mindset. My tree friend also empowered me to be more confident, able, and mentally and emotionally stronger. I plan on visiting my tree friend after this course. She has become a great friend to me and I plan on keeping this bond alive. If my tree friend and I were to fall in the world we could only make a sound if someone else is present.*
>
> *(Student 13, Montclair State University)*

Empathy and equanimity play a strong role in determining how students respond to individual and other perceptions and life's social and environmental challenges. There are a lot of challenges in the world that will push student responses to learning about life in objective and subjective terms. That is how the world works. Learning how to navigate that world is challenging for emerging adults that are just beginning to sow their own seeds and plant their own roots.

Providing the time and space for students to reflect on developing equanimity and empathy requires higher education to implement pedagogical practices that align with the spiritual, social, emotional and ecoliterate development of college students in terms of meaning making, purpose and a sustainability mindset that promotes generativity with them upon graduation. This sentiment is expressed in the Tree Journal Entry of Student 14, Montclair State University when they stated:

> *While Effie knows her future and fate, she has accepted her own journey. In the future, she knows she cannot do much for herself or even her tree friends, but she vows to make the best of the current life she lives. Effie hopes that, by the hands of Mother Nature, her life will be avenged by her own offspring when the time of her death comes. She says she sees the beauty of her own life and the things she's gotten to witness over her time, but she mainly hopes that the tree friends after her will have an even more fruitful life, filled with love and prosperity.*
>
> *(Student 14, Montclair State University)*

**FIGURE 7.6** Student 1, Montclair State University: Tree Journal. Photograph by Michael Lees

The further development of spiritual pedagogy and a sustainability mindset grounded in an ecoliterate-based learning environment aids students and educators in defining meaning and purpose in the work they are doing together. Creating an environment where asking personal and worldly questions, while developing student equanimity and empathy through the use of an exercise like the Tree Journal, supports student motivations while on their journey through higher education.

The second to last Tree Journal entry asks students to reflect on death, dying and change as the semester is also beginning to wind down. The questions address change, loss, impermanence and the nature of the unexpected happening in life. Working with change, loss, impermanence and the unexpected is no easy task for a human being of any age. Having the opportunity to reflect on impermanence provides the ability to open the mind and heart and apply meaning to the nature and quality of engaging in the present moment. Embracing what is learned, applying what is learned and doing life with the time that you have, teaches students to step into what they are choosing to do with their lives and an academic career. In closing I leave you with another of Student 1, Montclair State University's Tree Journal entries. This entry addressed life's nature and quality of change relative to personal, local, and global environments (Figure 7.6).

## Note

1 Throughout this chapter you will find supplemental learner artifacts that have been held onto throughout the years I have been using this exercise in the classroom. The learner artifacts are representative of the use of a journaling exercise relative to a spiritual, ecoliterate and sustainability mindset pedagogical approach to learning.

## References


Anderson, L. (1991). *Sisters of the Earth.* New York: Vintage.

Astin, A.W., Astin, H.S., & Lindholm, J.A. (2011). *Cultivating the Spirit: How College Can Enhance Student's Inner Lives.* San Francisco, CA: Jossey-Bass.

Blaser, R. (1993). *The Holy Forest.* Toronto, CA: Coach House Press.

Cervantes, M. (1605/1957). *Don Quixote.* New York: Mentor Books.

Chickering, A.W., Dalton, J.C., & Stamm, L. (2006). *Encouraging Authenticity & Spirituality in Higher Education.* San Francisco, CA: Jossey-Bass.

French, P.A., & Brown, C. (1987). *Puzzles Paradoxes and Problems: A Reader for Introductory Philosophy.* New York: St. Martins Press.

Hugonnier, B. (2007). Globalization and education: Can the world meet the challenge? In Suarez-Orozco, M.M. (Ed.), *Learning in the Global Era: International Perspectives on Globalization and Education* (pp. 137–158). Berkeley, CA: University of California Press.

Kassel, K., Rimanoczy, I., & Mitchell, S.F. (2016). The sustainable mindset: Connecting being, thinking, and doing in management education. In *Academy of Management Proceedings, 2016*(1), 16659. Briarcliff Manor, NY: Academy of Management.

Lees, M.V. (2017). Effect of contemplative pedagogy on the ecoliteracy of undergraduate public state university students. Doctoral Dissertation. Walden University.

Mansilla, V.B., & Gardner, H. (2007). From teaching globalization to nurturing global consciousness. In Suarez-Orozco, M.M. (Ed.), *Learning in the Global Era: International Perspectives on Globalization and Education* (pp. 47–67). Berkeley, CA: University of California Press.

Nash, R.J., & Murray, M.C. (2010). *Helping College Students Find Purpose: The Campus Guide to Meaning-making.* San Francisco, CA: Jossey-Bass.

Palmer, P.J. (1993). *To Know as We Are Known: Education as a Spiritual Journey*. New York: Harper Collins.
Robinson, K. (2011). *Out of Our Minds: Learning to be Creative*. Westford, MA: Capstone.
Safranski, R. (2005). *How Much Globalization Can We Bear?* Malden, MA: Polity.
Snyder, G. (1964). *Turtle Island*. New York: New Directions Paperbook.
Welch, M., & Koth, K. (2013). A metatheory of spiritual formation through service-learning in higher education. *Journal of College Student Development, 54*(6), 612–627.

## Appendix 1: Tree Journal

**If a You Fell in the World**
**Would You Make a Sound?**

### *A Semester with a Tree*

Welcome to your Tree Journal. Over the duration of the semester, you will find and meet a tree in a space and place on campus of your choosing.

This Tree Journal will involve inner and outer perceptions, reflections, and contemplations relating to class dynamics as well as your own life happenings.

The Tree Journal is a time for you to reflect upon and share with someone else, a someone else whom just so happens to be a tree. So meet and greet your tree, listen to your tree's voice and listen to what that tree may share with you.

The question-reflections for each of the journal entries serve as a guide for your thoughts and ideas. Use the questions found in each section as a tool to work with. You are not limited to only those questions, as you may want to add your own ideas as well.

If you want to respond to the questions in the form of a poem, writing, sketch, work of art, or a song feel free!

These entries will be a part of the final project and supplement your main journal for this course so make sure to keep up with the journal and most of all have fun befriending a tree!

### *Tree Journal Entry One – Week One*

**Voices**: What relations can hold between the real world and fictional worlds? Can real people interact with fictional characters? We are likely to feel that fictional worlds are insulated or isolated, in some peculiar way, from the real world, that there is a logical or metaphysical barrier between them. That, indeed, is why we call them different "worlds." – Kendall L. Walton (in French & Brown, 1987)

### *Question Reflections for Journal Entry:*

- Find a tree on campus that you would like to spend the semester hanging out with, introduce yourself, and draw a picture of this newfound tree friend for next class.
- What is your tree friend's name?
- How did your tree friend get here?
- What is your tree friend's background, life story?
- What does your tree friend have to do?

### *Tree Journal Two – Week 3*

**Voices**: Changing – Barbara Meyn (in Anderson, 1991)

It happens quietly. A maple seed blown here by a sudden random wind sprouts beneath the bedroom wall, grows before I quite know how it grew, tops the eaves, seeking afternoon as well as morning sun, and fills my life.

Leaves unfold like ragged green umbrellas waiting for an April rain. I tell myself it's just another tree that could have been dug up when it was small and planted farther from the house.

If I don't cut it soon, if I keep on watching while it reaches for the sky, delighting in its gray, sinewy trunk, the soft touch of leaves when I walk by, the way it gathers light on winter days and pours it generously through the glass, it won't be long until it moves my house off its foundation.

The room is full of curious, precious things, skin of mole, hawk feathers, moth cocoons, deer's-foot rattle, dry seed pods of zygadene, racemes of saxifrage. And now across the walls maple leaves sign to me in shadows.

Though the tree is not yet in the room, in the dark I hear it whisper, I know it's coming in.

### *Question Reflections for Journal Entry*

- What does your tree friend like to do?
- What is your tree friend's attitude towards life?
- What is your tree friend's philosophy on life?

### *Tree Journal Three – Week 5*

**Voices**: Breaklight – Lucille Clifton (in Anderson, 1991)

Light keeps on breaking. i keep knowing the language of other nations. i keep hearing tree talk water words and i keep knowing what they mean. and light just keeps breaking. last night the fears of my mother came knocking and when i opened the door they tried to explain themselves and i understood everything they said.

### *Question Reflections for Journal Entry*

- How does your tree friend think about other beings? People?
- Does your tree friend have a lot of friends or just some friends? Why?
- Does your tree friend trust the world surrounding it? Why or Why Not?
- Does your tree friend enjoy the companionship of a loved one? Friends? Family?
- Or does your tree friend like to be alone?

### *Tree Journal Entry Four – Week Seven*

**Voices**: Interlunar Thoughts – Robin Blaser (1993)

"Advertising tells us who we are" and "presents a completely integrated culture" in the interval between the old moon and the new when the moon is invisible, one hopes the moon will show up: capitalism, racism, consumerism, homophobia, sexism – all of them systems of signifiers detached from spirit so the governing soul goes numb (a voice on CBC set alongside John Wilkinson on John Wieners)

### *Question Reflections for Journal Entry*

- Does your tree friend have a religious, spiritual, atheistic view in relationship to the world-at-large?
- What is your tree friend's ideal vision of a relationship to self and others?
- If your tree friend could create the ideal world … what would that world look like?
- Does your tree friend believe in a sense of order as it pertains to relationships, life, and living, or does your tree friend just work with life as it comes?

### *Tree Journal Entry Five – Week Nine*

**Voices**: Our ideas can enslave or liberate us. Some people never do make the transition and remain resident in the old world view: their ideological comfort zone. In human life, there is always something new, because creativity is part of what it is to be human. – Ken Robinson (2011)

### *Question Reflections for Journal Entry*

- How does your tree friend feel about change? Does your tree friend feel like the world is changing all of the time?
- How does your tree friend think about itself in relationship to the space, place, and life on the planet?
- How does your tree friend think about itself in relationship to living on a planet that is part of universe?
- Does your tree friend believe that there is other life in the universe?

### *Tree Journal Entry Six – Week 10*

**Voices:** *Don Quixote's epitaph*: Here lies the noble fearless knight, whose valour rose to such a height; when Death at last had struck him down, his was the victory and renown. He reck'd the world of little prize, and was a bugbear in men's eyes; but had the fortune in his age to live a fool and die a sage. – Cervantes (1605/1957)

### *Question Reflections for Journal Entry*

- How does your tree friend feel about death and dying?
- Does your tree friend believe in life after death?
- How does your tree friend work with the idea: always expect the unexpected?
- How would your tree friend feel if you were gone?
- How would you feel if you went to visit your tree friend tomorrow and your tree friend was gone?

## *Tree Journal Entry Seven – Week 11*

**Voices**: The Uses of Light – Gary Snyder (1964)

It warms my bones say the stones I take it into me and grow say the trees leaves above roots below A vast vague white draws me out of the night says the moth in his flight – some things I smell some things I hear and I see things move says the deer – a high tower on a wide plain. If you climb up one floor you'll see a thousand miles more.

## *Question Reflections for Journal Entry*

- How was spending a semester with your tree friend
- What did your tree friend teach you and what did you teach your tree friend about life?
- Do you plan to ever visit your tree friend again after this course comes to an end?
- If you and the tree fell in the world would you make a sound?

***Note:** If you use this Tree Journal in your classroom please feel free to reach out and contact the author.

# 8

# "A STEP FORWARD"

## Developing a sustainability mindset among business students using a norm-critical approach and an adaptation of the "privilege walk"

*Lasse Lychnell and Karol Vieker*

> *"I will start being more courageous and dare to stand up for what I believe in. Because if I—who belong to the most privileged clique on earth—don't raise my voice for what is important, then who will?"*[1]

### Introduction

Meeting the United Nations' Sustainable Development Goals[2] will be difficult if the solutions originate from the same mindset that created them. Changing mindsets will be particularly difficult when the old mindset is governed by invisible norms that people may not even be aware of, such as the attitudes and treatment one receives based purely on skin color, gender or nationality. Thus, in order to develop responsible leaders, it is crucial to help students becoming aware of this dynamic and how their own taken-for-granted beliefs may help reproduce and reinforce a situation they actually want to change. This chapter introduces "A Step Forward," an adaptation of the "Privilege Walk" exercise (see, for example, Welton, Harris, La Londe & Moyer, 2015) with the purpose of developing insight into how norms affect the possibilities of achieving gender equality, reduced inequalities and equity on both an individual and a structural level. Originally used within the civil rights movement, here the exercise has been adapted to fit into the context of an international audience at a prestigious business school. It is based largely on the "A Step Forward" exercise that appears in the third edition of *BRYT! Ett metodmaterial om normer i allmänhet och heteronormen i synnerhet* (BREAK THE NORM! Methods for studying norms in general and the heteronorm in particular) (Åkerlund, 2011). In the exercise, students are randomly assigned roles that represent broad diversity in terms of, for example, gender, ethnicity, sexual orientation and socioeconomic class. During the exercise and the subsequent discussion, students come to feel and understand what it is like to be born with or without certain privileges and how that affects their opportunities in life.

DOI: 10.4324/9781003229735-9

## Background and context

Vital to meeting the Global Goals is the development of a mindset that facilitates the great changes necessary on different levels: individual, economical, societal and planetary (Rimanoczy, 2013). A mindset is the implicit lens with which an individual views the world and their place within it, and it rests on underlying assumptions, beliefs, and values that are rarely made explicit (Kassel, Rimanoczy & Mitchell, 2018). Every individual is, however, embedded in larger social spheres that are governed by intersubjective norms (Berger & Luckmann, 1966/1971), such as the family, the workplace and society at large. These norms keep society in a status quo for better and for worse. Consequently, changing one's mindset may be difficult even for those who actually want to. The aim of this chapter is not to account for the complex relationships between individuals' assumptions and social norms, but rather to present an exercise that helps make visible the extent to which norms govern our lives, as well as their consequences.

The exercise was used within the Global Challenges track[3] in the BSc Business and Economics program at the Stockholm School of Economics (SSE), a prestigious, privately funded business school that was founded in 1909. The aim of the Global Challenges track is to broadly educate students about urgent and large-scale problems of modern times, and it consists of four mandatory courses based on the United Nations' Sustainable Development Goals comprising in total 15 ECTS. One of the four courses called "Being" is particularly focused on questions like "Who am I in relation to the global challenges?," "What can and do I want to do?" and "How can I lead for sustainability, while at the same time being sustainable myself?"

During the seminar where the exercise is used, particular attention is paid to Goal 5: Gender Equality, and Goal 10: Reduced Inequalities. Thus, the purpose of the seminar is for students to gain insight into how norms affect the opportunity for gender equality, equity and fewer injustices on both the individual and structural levels, as well as how so-called norm-criticism can be used as a tool to counter these injustices.

This is particularly important as SSE students have traditionally come from privileged backgrounds and after their studies often go on to influential positions in private enterprise or politics. This gives them a unique possibility to choose to use their positions of power to affect positive change both in their own organizations and in society at large.

## Norm-criticism as a tool for change

Norm-criticism (Lövkrona & Rejmer, 2016) is a tool that can be used to reveal, review and change repressive structures that limit the rights and opportunities of individuals and groups that are perceived to be outside of what is considered to be "normal." It can be described as the opposite of tolerance, which does not take power structures into account. A tolerance-based approach does not answer the questions of who has the right to tolerate whom and what happens if they are or aren't tolerated.

For example, a tolerance-based approach may feel sympathy for people with a foreign background who are disadvantaged by the structural discrimination in the job market,[4] but will do little or nothing to change that structure. Furthermore, the responsibility of individuals to change repressive structures is not taken into consideration in a tolerance-based approach. Using norm-criticism, the perceptions of who is considered to be competent based solely on factors such as where they are from or what their name is will be critically examined, as will the implications of systematically rejecting applicants based on factors other than their competency or work experience. In doing so, the effects of norms related to background and competency are exposed as being advantageous for some but disadvantageous for others, from which a meaningful discussion around who is considered to be competent in various fields and how competency is assessed can ensue. This in turn can lead to measures being taken to reduce bias in recruitment, such as anonymizing applications and unconscious bias training for recruitment teams.

In summary, a norm-critical approach means:

1. Exposing norms and power structures
2. Studying the consequences of them
3. Showing who gains/ loses from following/ breaking them
4. Determining how we all as individuals contribute to reinforcing or weakening them, thereby empowering – and placing responsibility on – everyone to take concrete measures.

Because norm-criticism focuses on how all individuals contribute to creating and maintaining structures, it is an invaluable tool in bringing about the long-term structural change that is necessary to create and maintain open and inclusive environments.

A common misconception regarding norm-criticism is that the goal is to do away with all norms. Nothing could be further from the truth. Norms are necessary to, for example, facilitate communication (language norms) and prevent traffic accidents (norms regarding which side of the street to drive on). A context without norms would be characterized by chaos and anarchy. The goal of norm-criticism is rather, as the name implies, to critically review norms in order to determine which of them contribute to inclusion and which to exclusion. Actions can then be taken to reinforce the inclusive norms and weaken the exclusive norms, thereby contributing to achieving gender equality and reducing inequalities, for example.

## The exercise and students' immediate reactions

The opportunities someone has for work, housing and education vary greatly depending on, for example, their skin color, gender, sexuality, what socioeconomic class they come from and what functional capacity they have. The purpose of the "A Step Forward" exercise is to gain insight into the injustices that surround the advantages and disadvantages that individuals and groups can derive from following or breaking various norms. During the exercise, they do not act as themselves but rather take on the role of someone who may be similar to them, but who is likely vastly different from them. This is an adaptation from the original "Privilege Walk" exercise where participants act as themselves. While this is also an option, it is more effective to have students take on different roles in a fairly homogenous and privileged environment such as SSE. We also believe that it makes it easier for students to engage in the exercise when they play a role rather than act as themselves. The next section outlines the exercise (while detailed instructions are given in the Appendix).

## Short description of "A Step Forward"

At the beginning of the exercise, students receive a role card with a very brief description of the person they will be during the exercise, such as *You are a refugee who lives in hiding. You live with your family in one room in an apartment*; *You are a heterosexual mother of two children. You sell office supplies and live in a house*; or *You are 17 years old and study social sciences at high school. You have recently become a parent.* Students then fill in additional information about the person themselves, such as what gender they have, how old they are, what kind of family they have and what they do all day. The students then stand in a horizontal row and ten statements are read. If the statement applies to them, they take a step forward. Otherwise, they stand where they are. For example:

"I can buy skin-colored bandages that are similar to my skin color."

Light-colored skin? Take a step forward.

Darker-colored skin? Stay where you are.

In advance, the instructor chooses ten statements from the complete list of statements that raise the desired perspectives, for example:

- My native language, my religion and my culture are respected in the society I live in
- I can afford to go to the movies or go out for coffee once a week
- Every day I can read in the newspaper about successful people with the same skin color that I have
- I have never been worried that my salary is lower than my colleagues' simply because of my gender
- I have never had to tell my relatives what my sexual orientation is.

After the ten statements are read, the students are very spread out with some having been able to take all or nearly all of the steps, some having taken many of the steps, and some having taken few or no steps at all. Students are than asked to take careful note of where they stood when the exercise ended. They then gather in small groups with those closest to them to describe who they were during the exercise and what steps they took or didn't take. After then sharing a few examples in the large group, they are asked to return to where they stood at the end of the exercise. They are then asked to look forward and take note of what they see from where they are standing from a societal perspective (that is, not what they actually see in the room).

## Post-exercise discussion

Following the exercise, which requires a large space, the students reconvene in the classroom to discuss the exercise. The following questions are discussed in small groups and then in the large group:

1. How did it feel to take a step forward? How did it feel to stay where you were?
2. What happens when you end up at the very front or the very back? What do you see?
3. How can we go forward once we have learned about the advantages and disadvantages different people have in society? What can we do to counteract these injustices?

The first question, which focuses on the individual perspective, raises very strong feelings, particularly from those who ended up at or near the front or back or the group. Those in the front often express both feelings of happiness that things are going so well for them and feelings of guilt that others are being left behind or do not have their same opportunities or privileges. Those in the back often express both frustration and sadness over seeing how, no matter how hard they try, others are pulling away from them – and may not even notice it. This is typically expressed in their body language during the exercise as well, where their faces often fall and their shoulders sag. Some even express anger over the injustices to which they are being subjected.

The second question, which focuses on the societal perspective, challenges students to become aware of how individuals are advantaged or disadvantaged by the norms and power structures in a society. Here, students in the front often reflect upon how they see only possibilities from where they are standing, while students in the back reflect upon how they see barriers to the progress they so desperately want to make. As with the first question, students in the middle often express how they are aware that some people have more advantages than they do, but that they are relieved that they at least are not in the most disadvantageous position.

The final – and in the long run most important – question challenges students to use their newfound knowledge and insights to begin to solve the structural problems that have such a profound effect on so many individuals. These discussions both further emphasize the purpose of the exercise (to gain insight into the injustices that surround the advantages and disadvantages that individuals and groups can derive from following or breaking various norms) and serve as inspiration for the written assignment that follows the seminar where the exercise is performed.

## Students' written reflections after the exercise

Following the seminar and the "A Step Forward" exercise, the students are asked to perform the following written reflection exercise that is aimed at helping them take action based on the experience they have had:

> The norm-critical perspective teaches us that what we do on an individual level affects what happens on the structural level, no matter how large or small our actions may seem to be. We are all probably already doing things that contribute to equality and inclusion. There are also plenty of things that we could start doing. And if we're truly honest with ourselves, there are things that we should stop doing as well.
>
> Based on what you know about norms and the injustices different people experience in society, write down at least three things you will <u>start</u> doing, three you will <u>continue</u> doing, and three you will <u>stop</u> doing. They can be large or small, but be specific and motivate why you have chosen these things.

There was no specified minimum or maximum length of responses. In order to motivate students, they received one course credit (out of one hundred) for participating in the "A Step Forward" exercise and one for submitting the reflection assignment. This was also a conscious choice in the design of the course to create as non-judgmental an environment as possible and, thereby, make it easier for students with divergent attitudes to participate. Over two years, we received 309 responses, which makes a response rate of approximately 65 percent.

As with all student work, some students plainly put little or no effort into their responses, but the majority had obviously made a significant attempt to challenge their way of thinking and a few stood out as having put extraordinary effort into their reflections. Here we provide a selection of the students' own words that demonstrate an understanding of the underlying concepts on which the norm-critical perspective is built, as well as self-reflection around the implications of their own positions of privilege and power.

## I will start ...

> *... asking company representatives hard questions about why their values of diversity are not reflected in their leading positions, regarding both genders and races, since this is often the case.*
>
> *... appreciating in what aspect I've been given a "head start" so that I can better identify the injustice when other people are not being given the same opportunity.*
>
> *... listening more actively to different people's perspectives and backgrounds to better understand how they experience inequality and thereby gain further understanding of how my actions and inactions may affect them.*
>
> *... getting better at appreciating how privileged I am and how lucky I am to have the opportunities that I have – being more grateful to my family and friends that have given me said opportunities. I need to do this because it will make me humbler and more eager to help others get the opportunities I have always taken for granted as well as to understand their situation better.*
>
> *... to fully appreciate all the opportunities that I have been given so that I can see when one of these opportunities is not available to other people.*
>
> *... to question my own thoughts on what is normal/abnormal. For example, if I see someone who looks "different" (in relation to myself), then I'll stop and reflect on why I think that way, why I think that person stands out and so on.*
>
> *... to realize, not take for granted, and be grateful for the position and standing I have in society. How much of our success is due to our own achievements and how much is due to innate privilege is difficult to discern. Anyway, I know that, compared to many others, I have a greater proportion of innate privileges and thus have to use this head start in a wise and kind-hearted way.*
>
> *... thinking about how lucky I am when I'm having a hard time. A difficult exam period is nothing compared to being brought up in a war.*
>
> *... to learn more about political opinions other than my own and try to develop my understanding of why they suit other people better than they do me.*
>
> *... putting myself in my fellow humans' shoes. A useful exercise that hopefully affects how I treat others, but also something that creates a little self-insight into how good I have it as a white man and a student at the Stockholm School of Economics.*

## I will continue ...

*... to think about how privileged I am who can fill my stomach several times a day and drink and wash myself in clean water, and be careful not to waste more food or water than what is necessary.*

*... to read news from various newspapers. I think that what we read also shapes our view of the world. It is easy to only stick to newspapers or writers that confirm your own opinions, but I think it is very important to expose ourselves to different perspectives. This also helps us to stay open-minded about all types of issues.*

*... to think critically about my own behavior, principles and opinions – "Why do I think what I think and why do I do what I do?"*

*... to constantly remind myself not to be dragged into the standards of success and career choices that prevail at school. I often find myself comparing myself to others; I feel worse when I do not get as high grades, when I have to answer that I still do not know what I want to work with in the future or what ambitions I have. Just because I started at SSE, my values shouldn't have to change just because it feels like the surroundings say so.*

*... to be open to the ideas, opinions and ideologies of others. Although I have relatively strong personal opinions, it is always interesting and useful to listen to alternative viewpoints as well.*

*... to think critically about my values, opinions and behavior, "Why do I do what I do? Think what I do? Is it right? Is it fair? Is it productive?"*

*... striving for a career that gives me enough influence and resources to really make a difference and impact as many lives as possible for the better.*

*... being accepting of others' thoughts and opinions, even if they are different from my own. Personally, I feel that I am an open person who has an easy time taking in what others are saying. When I hear something from a perspective that I have not seen before, I always think through it before I formulate or express my own opinion on the thoughts in question. You gain a deeper understanding of each other through discussions about how and why you think differently. It is through consensus that people learn to accept each other, in my opinion. From personal experience, intolerance towards others can in many cases be traced to a lack of perspective.*

*... to maintain my gratitude. If I forget what privileges I have in life, that I am healthy, have two parents, a fiancée, top education, etc., then I will start to take all of this for granted. When my life becomes a right instead of gratitude, I forget to see my fellow human beings as equals.*

## I will stop ...

*... most often looking forward. Sometimes you have to turn around to see if there are those who you can help with a simple push from your position so that they can move forward.*

*... thinking that some "groups" of people have different ownerships of different issues. Gender equality, racism or anything alike are not issues that someone has more responsibility in solving than someone else. It is easy to find gender equality to be a battle that women are responsible in pushing forward. It is not!*

*... judging someone for being slightly chubby. Not everyone can or should reach an "ideal weight" and what is an "ideal weight" anyway?*

*... raising my voice in arguments and discussions. It is not democratic nor intellectually honest to try to outvoice someone who disagrees with you.*

*... belittling my own experiences and not believing that others will take them seriously, or that it doesn't matter. Like others, I have been in situations where people say or do disrespectful things, but I find it especially hard talking about my experiences with racism towards myself. This is because I think people won't take it seriously or that I am seeking attention. By not speaking up, I am also not saying that it's not okay and that it is an issue.*

*... thinking about clothing and such superficial articles that bring forward a stereotype. My thoughts here were inspired by the example discussed in the seminar of a highly qualified Roma woman who ended up being rejected after job interviews due to her attire. As soon as she changed to a normal Western outfit, she got past the interview stage immediately. It made me realize that if I were in an interviewer's shoes, I would have a massive mental barrier to break through in order to evaluate her fairly.*

*... judging people based on appearance and profession. This is something that I am ashamed of, but the other week I started to speak English with a dark-skinned taxi driver, to which he answered me, "I was born in Sweden." Such preconceptions contribute to oppression, which leads to exclusion, discrimination and insulting treatment.*

*... being silent when someone is joking or insulting someone else because of sex, sexual orientation or the like and instead speak up. It is so easy to have the feeling that the discussions that are being conducted are not appropriate but then sit quietly without saying something because you are afraid of how the others will react when you do.*

*... being the first to speak. I am generally quite quick to take the floor and therefore I contribute to the fact that as a large man I am perceived as more "authoritarian" than I need to be.*

*... taking for granted that I have good health and identify more with the situation of others whose everyday life is marked by the fact that their health is not tip-top.*

*... being so intense in my jargon if it is not needed since I am a tall and physically large man. I know that I can be perceived as threatening and dominant, which I really am not or trying to be. I don't want a physically smaller and less-confident person to hesitate in having a productive discussion with me.*

*... complaining about things that I myself have the potential to solve. This is important for my own sake (complaining is not good for mental health), but it also recognizes that other less-privileged groups have problems that are more serious than my own.*

*... interrupting people when they talk. Sometimes when I suddenly get ideas I tend to want to tell them right away without waiting for others to finish talking. It is not polite and can be perceived as if I do not like other people's ideas and just swamp them with my own.*

*... taking things for granted and being "blind" in my thinking. For those of us who already have it so good, it can sometimes be difficult to understand those who are actually worse off. In the "A Step Forward" exercise it becomes so extremely clear. The people in the back who are experiencing*

> *difficulties in society can see everyone else who is better off in front of them, while the people in the front only have a free path waiting for them without realizing that there are those behind them who want nothing more than to change their places every day.*

A number of students were expectedly provoked by the exercise, though they were clearly in the minority of those who chose to complete the assignment. For example, one student simply wrote, "*I will continue to believe in myself since I'm always right.*" Another wrote, *"I will start to help other people, in the silent majority, to dare to speak up. In academia for example, a lot of students fear actually drawing the wrong conclusions, writing the wrong essay or taking the wrong stance. This is not strange, since doing so, is often at the cost of your grade. Just look at this assignment – it is extremely left-leaning and biased in its entirety.*" Here it is worthy to again note that the assignment was graded based on whether or not reflection had been done, not on what the content of that reflection was.

As instructors, we choose not to place too much emphasis on comments like these. One of the purposes of the exercise – and indeed higher education – is to challenge students and their mindsets. Some students simply do not want to be challenged and we should not waste our valuable energy on them. All we can do is attempt to limit the space that they take in classroom discussions and, rather, focus on the majority of students who are interested in learning, reflecting and drawing their own conclusions.

## Discussion

While insights can arise in an instant, it may take years for behavior to change. The purpose of this exercise is primarily to give students an embodied experience of such insights, reinforced by their connection to other elements in the Being course, such as self-awareness, self-regulation, empathy, compassion and value-based self-leadership.

We have observed several immediate reactions among the students that point towards a mindset shift from the outside in, for example by triggering self-awareness, empathy and compassion, and awareness of their own values. This shift helps many students realize that they were born with certain privileges and those privileges have largely influenced their opportunities in life. Some students then feel a sense of responsibility to contribute to a sustainable world rather than only increasing shareholder value.

In the long run, one exercise may not necessarily change students' mindsets. However, like the first step in a classic 12-step program, it can help them to admit that there is a problem – that certain norms and power structures can and do have a negative effect on certain people's opportunities to reach their full potential. By becoming aware of this, those in a privileged position of power can use that position to, for example, achieve gender equality and reduce inequalities. We further believe that the experience from the exercise cannot be undone. Thus, later in life, certain situations may trigger this embodied experience and contribute to a shift in perspective which may help in generating more sustainable solutions.

The exercise described in this chapter is an example of a kind of teaching in which exercises are used to help students gain an embodied and subjective experience of the concepts that are taught. This approach suggests that we cannot just rely on "objective" knowledge without engaging with the phenomena with our own subjective experience in order to reflect from where our actions originate and the consequences they may have for other people, for the wider world, and for ourselves.

The strength of such an initiative is to a large extent determined by how well it is integrated into and supported by other elements, such as the course in which the seminar is given, the program as a whole, and the school's direction at large. SSE's overarching educational aims start with the idea that a successful future decision maker will approach the world with curiosity and confidence. As the philosopher Ingemar Hedenius said, they are "free and alive in relation to the unknown." Thus, the acronym FREE[5] has been chosen and students should be taught to be, among other things, Reflective (contemplate what role they play in the world and understand the impact they have on others), Empathetic (consider the perspective of others fully) and Responsible (seek improvement for society and the world at large).

The Global Challenges course's focus on the UN Sustainable Development Goals, as well as the norm-critical perspective and exercises such as "A Step Forward," reinforce SSE's educational aims of producing reflective, empathetic and responsible graduates. In short, as students are reminded at the end of the seminar:

- You have knowledge
- You have power
- You are privileged
- **You** can make a difference!

> *I am going to start reflecting on what I can do in my everyday life to reduce inequality in society today and realize the power I (as the privileged person I am) actually have to bring about change. It is not the people at the back of the ranks (in the "A Step Forward" exercise) who can achieve the greatest reform. Rather, it is we at the front who must open our eyes and realize that because of our situation in society, we can make the greatest difference together.*

## Notes

1 All student quotes are from students who have taken the Global Challenges course at the Stockholm School of Economics.
2 United Nations Department of Economic and Social Affairs (2020). Sustainable Development Goals Knowledge Platform. Retrieved from https://sustainabledevelopment.un.org/.
3 Stockholm School of Economics (2020). Global Challenges for the makers of the future. Retrieved from https://www.hhs.se/globalchallenges.
4 European Network Against Racism. Racism & Discrimination in Employment in Europe 2013-2017. Retrieved from https://ec.europa.eu/migrant-integration/librarydoc/enar-shadow-report-racism-discrimination-in-employment-in-europe-2013-2017 .
5 Stockholm School of Economics (2020). Mission and Vision. Retrieved from https://www.hhs.se/en/about-us/organization/mission-and-vision/.

## References

Åkerlund, Carl (Ed.). (2011). *BRYT! Ett metodmaterial om normer i allmänhet och heteronormen i synnerhet [BREAK THE NORM! Methods for Studying Norms in General and the Heteronorm in Particular]*. Stockholm: The Living History Forum and The Swedish Youth Federation for Lesbian, Gay, Bisexual, Transgender, Queer and Intersex Rights.

Berger, P.L., & Luckmann, T. (1966/1971). *The Social Construction of Reality: A Treatise in the Sociology of Knowledge*. New York: Anchor books.

Kassel, K., Rimanoczy, I., & Mitchell, S. (2018). A sustainability mindset model for management education. In K.Kassel and I.Rimanoczy (Eds). *Developing a Sustainability Mindset in Management Education* (pp. 3–37). London: Routledge.

Lövkrona, I., & Rejmer, A. (2016). Norm criticism: A method for working with core values. In T. Brage & I. Lövkrona (Eds.), *Core Values Work in Academia* (pp. 137–146). Lund, Sweden: Lund University.

Rimanoczy, I. (2013). *Big Bang Being: Developing the Sustainability Mindset*. Sheffield, UK: Greenleaf Publishing.

Welton, A.D., Harris, T.O., La Londe, P.G., & Moyer, R.T. (2015). Social justice education in a diverse classroom: Examining high school discussions about race, power, and privilege. *Equity & Excellence in Education, 48*(4), 549–570. doi:10.1080/10665684.2015.1083839

## Appendix: A Step Forward instructions and material

This exercise has been adapted for an international student body at a prestigious European business school. It is largely based on the version that appeared in the third edition of *BRYT! Ett metodmaterial om normer i allmänhet och heteronormen i synnerhet* (BREAK THE NORM! Methods for studying norms in general and the heteronorm in particular) as published in 2011 by the Living History Forum and The Swedish Youth Federation for Lesbian, Gay, Bisexual, Transgender, Queer and Intersex Rights (ISBN 978-91-977914-3-4).

The opportunities someone has for work, housing and education vary greatly depending on, for example, their skin color, gender, sexuality, what socioeconomic class they come from, and what functional capacity they have. The purpose of this exercise is to gain insight into the injustices that surround the advantages and disadvantages that individuals and groups can derive from following or breaking various norms.

### *Carrying out the exercise*

Make sure there are role cards for each participant and select ten statements in advance that raise the perspectives you want to address. Then give the following instructions to the participants:

1. You will receive a role card. You become this person during the exercise. Do not show the card to anyone else.
2. When you have read the card, fill in what is missing. What gender do you have? What kind of family do you have? How old are you? What is your ethnicity? What do you do all day? etc.
3. Stand in a horizontal row. Ten statements will then be read. If the statement is true for you (that is, the role you received), take a step forward. If is not true, stay where you are. You never go backwards.

   For example: "I can buy skin-colored bandages that are similar to my skin color."
   Light-colored skin? Take a step forward.
   Darker-colored skin? Stay where you are.
4. Read each statement, one at a time. Allow time after each statement for participants to make a decision about whether they should take a step forward or stay where they are.

5. When you have read all ten statements, ask everyone to carefully take note of where they are since they will later return to these positions.
6. Ask everyone to gather with those closest to them and tell each other who they were – both what was on the card and what they added, as well as what steps they took/didn't take.
7. Gather the group so everyone can hear and ask a few volunteers to tell the group who they were – both what was on the card and what they added, as well as what steps they took/didn't take. Also ask them to say a bit about how they made their decisions to take a step or not. Start with some who were in the middle, then some who were at the back, and then finally some who were at the front.
8. Finally, ask everyone to return to where they stood at the end of the exercise. Ask them to look straight ahead and notice what they see from where they are standing from a societal perspective (that is, not what they actually see in the room).

### *Discussing the exercise*

Once you have completed the exercise, discuss the following with the participants. They should preferably discuss in small groups before having a large-group discussion.

- How did it feel to take a step forward? How did it feel to stay where you were?
- What happens when you end up at the very front or the very back? What do you see?
- How can we go forward once we have learned about the advantages and disadvantages different people have in society? What can we do to counteract these injustices?

### *Statements*

- My religious holidays are also public holidays.
- No one has ever asked me if I am a man or a woman.
- I don't need to save money at the end of the month.
- No one has ever said I'm in a bad mood because I'm menstruating.
- I am not worried about my future.
- My possibilities of getting a job are not negatively affected by my name.
- My native language, my religion and my culture are respected in the society I live in.
- I am not afraid of being stopped by the police.
- I can buy skin-colored bandages that are similar to my skin color.
- I can travel wherever I want to without having to find out in advance if there is assistance for the disabled.
- I have never been called something derogatory because of my sexual orientation or my gender expression.
- I can go swimming in a public pool without having to think about which changing room I should use.
- I have never been called something derogatory because of my skin color or where people think I'm from.
- I can wear whatever clothes I want to.
- I can choose to work with what I want to.
- My voice is often heard when we take breaks at work.

- I can afford to go to the movies or go out for coffee once a week.
- No one has ever looked at me suspiciously while I've walked down the street.
- No one has ever asked me where I'm really from.
- People assume I'm good with computers simply because of my gender.
- I have not had any financial difficulties.
- I have an EU passport.
- I can easily enter a building even if there are stairs up to the door.
- I have never been ashamed of my home or my clothes.
- No politician would ever question whether I am suitable to be a parent.
- I feel that people listen to me and take my opinions seriously.
- I can visit a government authority without needing an interpreter.
- I can walk down the street holding hands with the person I love without getting strange looks.
- Every day I can read in the newspaper about successful people with the same skin color that I have.
- If I were visiting the parliament, people would think I was a parliamentarian.
- I don't have to look for reduced prices when I buy food.
- My parents and teachers have made me feel like I can be anything I want to be.
- I am rarely afraid if I'm out in the city at night.
- It is easy for me to get information written in my native language.
- I live in an area with relatively low unemployment.
- No one has ever called me "sweetheart" in a professional setting.
- I don't have to be on my guard if I walk past a group of men late at night.
- I have never been worried that my salary is lower than my colleagues' simply because of my gender.
- I have never had to tell my relatives what my sexual orientation is.

## *Roles*

- You are a 21-year-old woman of indigenous heritage. You work in an office and use a wheelchair.
- You are a 30-year-old Jewish man. You are an actor but work as a nursing assistant.
- You are a 30-year-old woman. You study at a university and have lived in a new country for five years.
- You are married and sit on the municipal council. You have three children and a nanny. You have dyslexia.
- You are the president of a political youth organization. Your parents came from Chile during the 1970s.
- You are a young man who lives in the suburbs of a large city. Your mother works as a cleaner and your father is unemployed.
- You are a young Muslim woman. You live with your parents who are very religious. You are studying to be a lawyer.
- You are a heterosexual mother of two children. You sell office supplies and live in a house.
- You are a 17-year-old woman who didn't finish school. You work in a fast-food restaurant.

- You were born a girl but ever since preschool have felt that you really are a boy. Your parents are teachers.
- You are a 15-year-old woman. You live at home with your parents in a small village. You are adopted.
- You are a heterosexual man who works as a dancer at the opera. You grew up in a small community.
- You are a 50-year-old man. You work as a police officer. You have recently divorced your husband.
- You are 17 years old and study social sciences at high school. You have recently become a parent.
- You are the daughter of an assistant nurse and study economics at a university. You play floorball in your spare time.
- You are a young man who has fled Iraq with your family. You are in 9th grade and dream of being a doctor.
- You are a 25-year-old white man who studies at the Stockholm School of Economics. You play tennis during your spare time.
- You are a refugee who lives in hiding. You live with your family in one room in an apartment.
- You are a woman who works in healthcare. You have lived in the country where you are for 14 years. You live with your girlfriend in a suburb.
- You are single and just over 30 years old. You are a high school teacher and still live in your hometown.
- You are the daughter of an American ambassador. You are Christian and live in Western Europe. You go to a school for the visually impaired.
- You are a 45-year-old woman of Roma heritage with an engineering degree. You have lived in the same country for ten years.

# PART II

# Action on the SDGs

# 9

# BECOMING A REFLECTIVE STORY-TELLER

## An empowering and transformative learning experience with AIM2Flourish

*Ekaterina Ivanova*

### Introduction

This chapter describes and discusses how AIM2Flourish as an innovative experiential learning tool can serve for developing a sustainability mindset (Naim, 2018; Nonet & Petrescu, 2018; Rimanoczy, 2017). According to the founding fathers and later advocates of experiential learning (Dewey, 1938; Kolb, 1984; Mezirow, 2009), giving students the opportunity to learn by doing or through experience is much more meaningful and has a more lasting impact on their holistic personal development than traditional lectures focused on remembering and learning the right answer. Being a foundation for life-long learning, experiential learning is grounded in constructivist theory, which recognizes sociohistorical origins of knowledge and its acquisition by means of social mediation (Vygotsky, 1978). By building on prior learning, students are constructing knowledge through social activities and processes guided by interaction with the outside world (Wertsch, 1985) and mediated through reflection (Mezirow & Associates, 1990). Reflective practice as a metacognitive activity plays a key role in experiential learning and is defined as "the process of bringing past events to a conscious level and of determining appropriate ways to think, feel, and behave in the future" (Caffarella & Barnett, 1994: 38). Experiential learning is considered transformative in the sense that it is "learning that transforms problematic frames of reference to make them more inclusive, discriminating, reflective, open, and emotionally able to change" (Mezirow, 2009: 22). Through such innovative approaches to learning, students are prepared to understand the complex problems of the modern VUCA (volatile, uncertain, complex and ambiguous) world and to explore multiple solutions to the wicked problems humanity currently faces.

AIM2Flourish as a transformative learning tool could be used by management educators to facilitate the sustainability mindset among business students on their way to becoming a new generation of globally responsible entrepreneurs and corporate leaders capable of making an impact toward a more sustainable planet (Fairfield, 2018). For this to happen students go through a guided process of transformation, so that the principles of the sustainability mindset become internalized within the three dimensional contexts of being (one's values, identity, emotions and creative elements connected mostly to the right hemisphere of the

DOI: 10.4324/9781003229735-11

brain), thinking (knowledge, cognitive and analytical capabilities linked to the left hemisphere of the brain), and doing (competencies and actions) (Kassel & Rimanoczy, 2018). This triad of interconnected components necessary for cultivating a sustainability mindset rests upon four components that could be intentionally developed: ecological awareness and systems thinking, as well as emotional intelligence and spiritual intelligence (Rimanoczy, 2021).

AIM2Flourish is a learning assignment that works perfectly with students at the personal level, and enables them to question some of the underlying assumptions about the sustainability of the world in the Anthropocene era. For instance, unlimited growth driven by the neo-liberal understanding of the purpose of business and of profit generation only for the sake of shareholders are examples of such assumptions. Another way of framing existing assumptions is by distinguishing between weak and strong sustainability. "Weak sustainability" is a term that assumes that unlimited economic growth is indispensable and decisions are made based on values of progress driven by economic motives and ever-increasing GDP. In contrast, "strong sustainability" is used to point to a future created without growth as a key measure of societal development, instead, decisions are driven by their impact on the environment and welfare for all living creatures (Landrum & Ohsowski, 2017). Reflection on the individual values that matter to students and guide their behavior after course completion is one of the key outcomes that this learning activity makes possible, so that students are prepared for the transition towards a no-growth society based on principles of a low carbon, circular and local living economy. Since change is caused by innovative thinking and perception of reality, the shift in students' mindsets depends on a holistic approach that requires an effort of self-discovery with an introspective focus on the deeper self, making a meaningful connection to the outside world, and taking action (Rimanoczy, 2017). Graduates who have had an opportunity for such critical reflection about their personal assumptions are proactive learners who tend to develop a better awareness of what they want to achieve in their professional lives (Billett, 2009).

## AIM2Flourish as an experiential learning tool

AIM2Flourish is considered to be one of the most innovative learning initiatives for developing activist business leaders, promoting responsible management education, and achieving sustainable development (Farias & Waddock, 2018; Storey, Killian & O'Regan, 2017: 99). Launched at the UN PRME Global Forum in 2015, AIM2Flourish is a program, hosted by the Fowler Center at Case Western Reserve University, that supports a global network of educators and students in disseminating the stories of businesses striving to implement the 17 UN Sustainable Development Goals (Buchter, 2018: 54). The AIM2Flourish program was created by the Fowler Center for Business as an Agent of World Benefit with the goal of shifting the worldview of students. As this chapter is being written, AIM2Flourish has reached more than 2,100 participating business schools, including 500 professors around the world and a global community of more than 9,300 members that spans 81 countries. More than 3,350 stories about business as a force for good have been published. By taking part in the AIM2Flourish initiative, future business leaders are guided towards understanding the 17 UN Sustainable Development Goals (SDGs) and how business can contribute to achieving them, and are inspired to take further action for building a flourishing world for all.

As the name of the Fowler Center suggests, students embarking on the AIM2Flourish journey are taught to see business outside of the narrow paradigm of "maximizing shareholder wealth." Instead, students are encouraged to apply the redefined purpose of business in society as a collective value creation (Donaldson & Walsh, 2015) and adopt a wider stakeholder-driven paradigm (Freeman, Parmar & Martin, 2020) grounded in systems perspective by applying such concepts as "flourishing enterprise" (Laszlo & Brown, 2014), "conscious capitalism" (Mackey & Sisodia, 2013) and "circular economy" (Webster, 2017). Thus, an underlying assumption is that companies that care about the flourishing of their communities, the dignity and well-being of their employees, creating real value for their consumers and leaving zero-footprint on the environment (Sternad, Kennelly, & Bradley, 2016) as a transformational strategy in line with the ideals of regenerative capitalism (Elkington, 2020) show better performance on their triple bottom line.

The key feature of AIM2flourish as an experiential learning assignment is outside of the classroom – in the engagement with a real business already involved in building a business model embedded with purposes based on the SDGs. Students are confronted with the task of connecting directly to the business of their choice, one that inspires them. They have to arrange an interview with a business leader of a sustainable business that acts as an agent of world benefit and maintains financial soundness at the same time. In order to follow the mission of AIM2Flourish "to change students' mindsets about the goal of business from being the best in the world to being the best for the world" (Buchter, 2018: 55), students use Appreciative Inquiry (Cooperrider, Whitney & Stavros, 2008) as an interviewing method. By using this approach, students are able to recognize the positive side of doing business. During the interview, students ask strength-based questions to uncover the moments in life when business leaders felt themselves at a high point – most engaged and passionate. Since they are making an inquiry into business innovation, they ask the leaders questions about the motivation and inspiration that led to this high point. Finally, students ask about the business, societal, and environmental impacts of the company. This interviewing approach naturally leads the interviewees to story-telling. As an outcome of such interaction, students listen to stories and convert them into online AIM2Flourish publications, and by doing so are inspired themselves to become future responsible business leaders (Buchter, 2018: 55).

## Embedding SDGs as a learning framework

AIM2Flourish was designed as the world's first higher-education program to integrate the 17 UN SDGs with the UN's 169 targets for 2030 into the learning process, in order to uncover the potential of business in addressing the wicked problems that humanity faces. The Global Agenda 2030 adopted in September 2015 by all 193 members of the United Nations represents an ambitious inclusive plan for achieving a better future for all. The Decade of Action started in 2020 calls for finding and implementing sustainable solutions to the biggest challenges in the world related to poverty, inequality, climate change, injustice and peace. The SDGs at the heart of the Agenda 2030 are constructed "to mobilize the creativity, know-how, and resources of multiple stakeholders to take global action" (Wersun et al., 2020: 9). Since the SDGs were initiated by the United Nations, working on achieving them became an explicit expectation for all business schools that are signatories of the United Nations Principles for Responsible Management Education (PRME), just as similar expectations are true for the members of the UN Global Compact (Fairfield, 2018). As management scholars focus on discovering the role of business in implementing this ambitious global agenda (Howard-Grenville et al., 2019), students taking part in AIM2Flourish are also making their contribution to addressing the SDGs. By learning about businesses through the SDGs framework, students are capable of transforming themselves into ethically aware and sustainability-literate future responsible leaders (Dean, Gibbons & Perkiss, 2018).

The powerful combination of the Appreciative Inquiry (Cooperrider et al., 2008) methodology with the SDGs serves as a guiding framework for students taking part in the AIM2Flourish project as they search for a business making a contribution to one or more of the 17 SDGs. Students in Russia have already discovered some of the astonishing conscious entrepreneurs and their sustainable business innovations, including the following:

- Yandex.Taxi, the most popular service for ordering a taxi online in Russia, has designed and implemented an innovation project adapting their business model to hearing-impaired and deaf drivers
- BioFoodLab became a pioneer company on the Russian market of healthy vegan snacks affordable for people of all ages and income
- LavkaLavka is a social enterprise that aims to revive agriculture and lost gastronomic traditions in Russia, supporting local farmers based on the farm cooperative business model
- Windows Factory's innovative approach is not just to run the business efficiently, but to deal with the company as a living organism, giving employees all the necessary conditions to flourish and develop themselves
- Zero Waste Shop, being the first of its kind in Moscow, refuses to use environmentally unfriendly plastic packaging and sells only organic and multiple-use products.

AIM2Flourish is an open-access global library of inspiring stories recorded and told by students about the ways that businesses set up for making money also serve the broader world by doing good. Each year, through two rounds of competition, the 70+ best stories are celebrated as Flourish prize finalists; of these 70+ stories, 17 of them will be declared Flourish prize winners – one for each UN SDG. The prizes recognize the accomplishments of the students as authors, the professor who supervised the story's publication, and the business leader interviewed for the story. Such an approach to publicly celebrating good practice helps to motivate and engage students, generates recognition for the businesses that opened their doors to cooperate with students, and adds global visibility to the business schools involved. As such, AIM2Flourish facilitates promotion of the SDGs in business schools and within the business community. In the six short years since its establishment, AIM2Flourish has been showing quick growth and operational sustainability based on the support of its hosting organization and a network of partners that enable the functioning of highly engaged staff and volunteers in an online-based platform setting. The partnership strategy of AIM2Flourish has become more formalized over time, which helps to maintain its focus on for-profit businesses, thus aligning with the shared vision of the founding partners, such as UN PRME, GRLI and Flourishing Leadership Institute (AIM2Flourish, 2021).

## The context, learning goals and details of the learning tool

The context for AIM2Flourish is an elective course, "Responsible Business and Sustainable Development." Since 2016, the course is offered annually at the Graduate School of Business of the HSE University in Moscow. The course takes place during the winter semester for second year students in master's degree programs in management and provides an opportunity to earn 4 ECTS units. On average, 40 students sign up for this class, with approximately 20 percent of international students and 80 percent of local students from all over Russia. AIM2Flourish is integrated as a final group assignment in this course with the following learning goals:

- To facilitate a transformation in students' mindsets towards accepting sustainable and responsible ways of doing business as a new norm
- To provide students with experiential learning opportunities, based on implementing a real-life project and reflecting on the interaction with responsible business leaders
- To empower students to develop stories for the AIM2Flourish platform of local "businesses for good" engaged in sustainable innovations – stories that address one or several of the UN SDGs.

To reach these goals, students are inspired and empowered through an outside-the-classroom experiential learning activity as they identify and interview a leader of a responsible and sustainable business for the AIM2Flourish platform. Students are assigned by the educator to work in a multi-cultural team setting consisting of Russian and international students. Together, they complete the task of finding a sustainable business innovation that their "hero" implements in order to contribute to one or several of the UN SDGs. Detailed instructions about the AIM2Flourish assignment are provided in the course syllabus in a four-page-long appendix. This syllabus outlines the value of experiential learning and how teamwork should be organized to meet the criteria of the AIM2Flourish platform (focusing on profitable businesses that address the real needs of the community that correspond to the relevant SDGs), and also gives clear guidance on grading, deadlines and workflow. To avoid possible misunderstandings by students selecting a nonprofit organization instead of a business, it is highly advisable for the educator to confirm the choice of business before the students initiate a contact with their case protagonist.

This paper's author served as the designer and educator responsible for managing the course content and assignments, arranging guest speakers and company visits. The class was delivered in an on-campus format in two blocks, with the first five sessions offered in September–October and the last five sessions in November–December. Such an approach gave students enough time to implement their AIM2Flourish project in order to present the project results in mid-December, during the last session. The timing was perfectly suited for the stories to participate in the annual competition for the AIM2Flourish Prizes. Using extensive guidelines and materials available on the AIM2Flourish platform, the final assignment for the course was designed to connect students with real business leaders. As mentioned before, students used Appreciative Inquiry (Cooperrider et al., 2008) to gather information and write stories about their heroes that highlight how their work aligns with the UN SDGs. Even though AIM2Flourish as a final group assignment corresponded to 30 percent of the final grade for the course, it is important to mention that to earn the remaining 70 percent of the grade, students were expected to show active participation in the in-class discussions, write reflection papers on a walk in nature exercise, attend sessions of guest speakers talking about the best CSR and sustainability practices in Russia, complete an in-class essay and make presentations on cases, assigned readings and videos for the course. Such an approach to course design provided students with an understanding of key concepts and practices underpinning the responsible way of doing business, enabling them to be fully prepared for the AIM2Flourish experience. Class activities, readings and pre- and post-AIM2Flourish assignments were designed to scaffold student learning about the SDGs and responsible business as a new norm. Throughout the course, students were asked to reflect on their learning in face-to-face classroom discussions and written assignments, both individually and collaboratively. Based on their AIM2Flourish experience, each student was required to write an individual reflection paper on teamwork in general, the interviewing process, interaction with the AIM2Flourish platform, mentoring received from their educator, and any other issues. The students' reflection papers included in this article were selected from a class taught in the winter semester of 2018. It was the third year of the author's engagement with the AIM2Flourish platform.

## Students' reflections on the AIM2Flourish experience

Full versions and excerpts from students' reflection papers are used here to illustrate the transformative power of the AIM2Flourish experience. The reflections are grouped based on four AIM2Flourish stories they co-created. Some of the reflective essays highlight all stages of the AIM2Flourish journey and the impact it had on the shift towards a sustainability mindset, others point out the most emotionally intensive moments that were transformational for the student. Two out of the four stories presented here became finalists for the 2019 Flourish Prizes:[1] "Enabling Russia's Digital Future" and "Equal Opportunities for Deaf Taxi Drivers." The students were all non-native English speakers, and the author decided to feature their reflections without editing, to give the reader the powerful experience of how these nine students express themselves in such spontaneous ways with the language tools they have.

### *AIM2Flourish story for SDGs 4, 8 & 17: Enabling Russia's digital future*[2]

*Reflection # 1, HSE University international student from Germany*

*In order to create our AIM2Flourish story, we had to think about many different companies that are successful nowadays, and meanwhile contributing all in a slightly different way to a sustainable business and future. We learned a lot about their history, about the way they started from scratch and managed to earn a sufficient amount of money for the living of the CEOs and the employees. We soon had to pick the company we will focus on for our AIM2Flourish story. Because we had to do an interview with the CEO of this company, it was favorable if it is located in Moscow or the surroundings. From here on, plans got more and more concrete. The by far the most interesting part of the project was organizing the interview with the CEO of the company. In our case, we spoke with the CEO of Kodabra [Daria Abramova]. She is still very young but managed to create a business that expands rather quickly economically and which also contributes to a good and sustainable development of young people in Russia. The company focuses on teaching programming to juveniles. The project started as a non-profit organization, and now turned into a profitable business.*

*As far as I am aware of, all the students of our course were born between the 1990s and the 2000s. We are the generation, that will have to focus fully on climate change measures and therefore onto sustainable businesses. What we mess up now and forget will highly influence and modify the lives of all the generations coming after us. The economic system most companies are pursuing nowadays will dramatically degrade our actual environmental situation. Changes have to be made if we want to avoid the worst-case scenario for our climate. AIM2Flourish helped me to understand, in what directions companies will have to go.*

*Resources are not infinite. Companies, which have no environmental footprint are the future. Especially politics will have to intervene more severely in the way companies operate. Right now, private projects like the B-Corp company are very nice and taking things in the right directions. But private incentives do often not suffice, especially if sustainability is costly for enterprises. We live in a capitalistic world, where profits are the only way to truly appease investors. If politics do not push the companies to the right directions, many will keep operating like they do right now, meanwhile harming the environment. AIM2Flourish proved to me, that a sustainable and a profitable business are two compatible things. And it will certainly motivate me to get companies in these directions, either in the economical or in the political sector. Right now, I cannot decide between the two. I will still have to wait in what direction my future takes me.*

*The problems of the 21 century do not only come down to environmental issues. Social issues are becoming more and more important. How much should employees earn? Do we need minimum wages? How can a modern family with children work and have a family live in parallel? How can woman be better integrated in leading positions? What about the general work/life balance issues? Our AIM2Flourish project was socially sustainable, not specifically environmentally. Juveniles in Russia get the possibility to improve in programming, a sector that will still dramatically increase in importance in the years coming. May be, these kids are the future of the Russian economy, which currently faces a lot of difficulties. AIM2Flourish showed me, how important sustainable social frameworks in companies are. Generally spoken, a happy employee is a more productive employee with less off-time. And if, besides productivity, his personal quality of life is increased due to a social sustainable path a company takes, we find ourselves in a win-win situation.*

*Sustainability should not be seen as a way to obtain more profitability and a positive image in front of the customer, but as a duty that we all have. And the course clearly showed us, as already said, that this model is future proof and compatible with a sustainable business. I hope that many more modern enterprises realize, that they are all part of this duty. It is not unlikely that some members of our courses will in the future occupy positions in politics and in economics, where they have the power to decide to what extend companies become more sustainable. I personally believe that the course gave all students the incentive to prioritize the sustainable goal in future projects over others, even if it may sometimes come down to a more expensive production costs, or to less sales. Most sustainable products last longer than cheap ones with low qualities. A longer lasting product will be replaced less often, which brings down the sales in a long-term perspective. But in a long-term perspective, we want to preserve our climate. So, in any way, it should be worth it.*

*Managing the AIM2Flourish project was not always easy. First of all, it can be rather hard to find a matching meeting date with a CEO that is always extremely busy. It took us several attempts, and when the meeting was finally fixed, we had to wait nearly an hour before our interview partner was available. Also, it was sometimes hard to distribute the tasks between the different team members, especially if teams are rather big, and composed out of seven people. Of course, in a company, teams are a lot bigger. But in a company, the team members are generally very focused on one single topic rather than on several different subjects with different schedules like it is the case in university. And also: people get paid, which makes them usually more motivated than some students who sometimes behave like free riders. Hereby I do not want to say that the composition of rather big teams was negative. It was an additional task that brought me forward and sharpened my organization skills.*

*One main part of the course, and the one I probably most loved, was the fact that we had to think so often about our daily actions. We might assume that this is normal and should be required in every course in university. But in the years and the courses I did, my experience was that personal reflection often came too short. Doing readings can sometimes indeed be hard for students. It is not always easy to get the motivation of every student to do long readings. But in addition to the readings, we had to watch interesting and entertaining videos and ask ourselves and others questions about our and their daily actions. These questions really made me think in different directions. This kind of learning was totally new for me, and I really appreciated it. AIM2Flourish helped me to answer many questions I had before in regards of start-ups, and especially in regards of sustainable start-ups that create by operating no significant ecological footprint and positive externalities. It helped me organize myself in a team, working together, accomplishing many tasks that require a lot of discipline.*

### *AIM2Flourish story for SDGs 7, 9, 12 & 13: The road to a sustainable future*[3]

*Reflection # 2, HSE University student from Russia*

*On the whole the project for AIM2Flourish was a huge challenge for me. I have spent so much energy and nerves on it I have never actually did on any other project. Firstly, we had a lot of difficulties with the teamwork. Personally, I believe that six people is way too much for any project. Three is an ideal number, and according to my experience the fourth member is always almost a free rider … I have spent way too much time worrying and being disappointed that there are people in this world that seem to not feel responsibility to their mates. The second challenge was using English language. I had quite a good level before and during this course I suddenly felt that I somehow had lost it. It was not a pleasant experience that I had to ask my friend to help us with revising our AIM2Flourish story, though I knew I could count on her completely.*

*The interview was certainly a good thing and I knew it would be. I just love conducting interviews, I did that a lot when I was getting my Bachelor's degree. Natalia [Malashenko, Stakeholder Relations Director of UPM in Russia] was such a nice interviewee with proper scheduling and trying to answer all our questions. She has signed a story publication agreement without any hesitation. I also liked working with our story and interacting with AIM2Flourish platform. It looks so nice and is very convenient to work with. I have also enjoyed editing our photos, making them to look bright and beautiful, and making our presentation. I really like doing all the visual stuff.*

*I would like to thank our Professor for guiding our work and all the mentoring. I felt like even if everything went wrong we still would have got support and somehow fixed it with the help of our tutor. That has made me feel safe even though the task was not really simple. All in all, I am satisfied with the results, though I am a kind of perfectionist and believe that it is always possible to do things better. I think, we have broaden our horizons, done a really good thing trying to tell people of the world of UPM company which may be not so famous as it is operating mostly in B2B sector. I hope our story will be published. As I have already told in "My legacy" essay [visualizing a future thank you speech as a Nobel Prize winner in literature: "No matter what, keep fighting! There are people who will miss your never written stories"], it would have been much worse, if this certain story had not found us.*

*Reflection # 3, HSE University student from Russia*

*"Crucial part of study was creation of AIM2Flourish project, as a practical part. First of all, experience of interactions with Natalia [Malashenko, Stakeholder Relations Director of UPM in Russia] was very positive. She replied to us very soon and was easy to access. The interview passed in a very polite, friendly and good atmosphere, Natalia was open to questions and mentioned many other interesting aspects, which made our project more diverse. To tell the truth, we even made several videos, but did not manage to make a good story out of it later. As for me, the main point, which actually motivated me to make a story nice and capturing, was the thought that we will have to submit our story not only to our teacher, so that, if made in a bad way, it would not be a really huge disaster, after all (though I must stress that the inspiration of the professor on the subject played a crucial role in the importance with which I treated the course), but the fact that we need to present it to other teams as well as send it to the platform for many people to see! That was what actually drove me during the preparation process, and that was where some stress appeared.*

*The course enlarged my views on the CSR itself. At the very start of the course, it seemed to me to be connected solely to business, as I have studied the similar course before. I was very surprised by the absolutely different approach, which suggested not only business, but our personal lives as well. At first, I felt a bit skeptical and defensive, but by the end of the course I felt in a much more positive way. Frankly speaking, I am a bit impressionable person, that is why I was highly involved in course ideas and thought a lot about them outside the class. I cannot say that I implemented sustainability into a lot of my daily activities (such as separate waste or refuse of plastics bags for waste); however, I am on the way of building much more conscious perception. What is important, I started to catch myself on thinking, whether something what I did make harm to other people or nature. The other crucial change that I experienced was the change of a company's perception as a place for my job. I would say that now I understand how much inspiration matters, and how ability to work not only for your own and company's profit, but also because it is important and enables to solve really significant issue – of nature or society – matters for the inspiration.*

### *AIM2Flourish story for SDGs 3, 8, 9, 11 & 12: Changing Russian recycling culture – mission impossible?*[4]

*Reflection # 4, HSE University international student from Germany*

*I really enjoyed the CSR course and it was even one of my favorites at the HSE during my semester abroad. First of all, it was a really unique seminar with unusual teaching methods for me but which were great. While having such an "open atmosphere" during the lecture where everyone is asked for his opinion, it makes much more fun to participate and also makes it easier to follow the content. Moreover, I really liked the guest speakers and especially that you manage it to provide us so many of them and from different areas. They gave me a completely different view at the topic. From the very beginning I was fascinated from the AIM2Flourish platform and the whole idea behind it. I have never heard about it before and I am really proud that now I can call myself a participant of the project and a part from all of the people from all around the world who work for it. Working on our story provided me with some great insights into a Russian company and their CSR. Moreover, I really enjoyed the stories from the other teams, throughout which we had the chance to experience not only our story but got the chance to follow also CSR of other companies, from very small one up to the Big Four, and the working process from the other students.*

*At the beginning, the idea of working in a group of seven seemed to be really tricky for me. I never worked in such a big group during my studies and I was really curious how we would manage it. It is not surprising that sometimes we had some troubles to manage the group work for different tasks and divide them the way that everyone would feel satisfied, but I never had a feeling that it is impossible to manage it or that there is someone not willing to do anything. Moreover, with the help of google doc and the great opportunity to work at the same document from different places we found a "common language" and managed it to work together efficiency. By looking back, I have quite a good feeling towards the whole group project. I never had a feeling that I am alone or won't get help or an answer from my group mates. Moreover, we reminded each other of our homework and other tasks which we had to do, so the group was a support during the whole semester not only for our group project. I also loved the idea that we had mixed groups with local and international students combined, so I could work with students from different countries and become more familiar with other cultures.*

*Throughout the lecture and with the help of the small tasks which we got as our homework I gradually started to change my behavior towards significant topics as for example my water and plastic consumption. Born and raised in Germany I was surrounded by sustainable life and therefore I kind of became also a part of it, e.g. I brought plastic bottles to the supermarket but to be honest it was based more on the money I got than that I cared about my environment. Participating in the CSR course helped me to become aware of my environment and finally helped me to understand why everyone has to care. I became aware of the tasks I have to fulfil and the environment about which I have to and also want to take care. I already started to convince people who surround me as well and I will definitely continue doing so. Finally, I am really thankful for the unique and very diverse course, which provides me not only with lots of theory but also with practical insights and made me a part of the AIM2Flourish platform.*

*Reflection # 5, HSE University international student from Belgium*

*I loved the AIM2Flourish project, because the goal and aim they want to achieve is really nice. The fact that we can share what we learn with a whole platform with people worldwide is amazing. The understanding of the platform is easy and user-friendly. I had no difficulties with writing the story and know how to submit it. When having some spare time, I often read some stories published on the platform to learn more about the different businesses that are socially responsible in some way. For me the AIM2Flourish assignment made me aware of how many businesses are trying to make an impact for society. I learned about appreciative inquiry interview and think it is a nice method to conduct an interview and give back more interesting results. It also stimulates the interviewee to think in a different way and this brings them sometimes out of their comfort zone which is nice.*

*The thing I liked the most about this class was to hear the opinions of everybody on some sustainable matters. The fact that there are a lot of people from different nationalities makes it even more interesting, because I realized there still is a cultural difference between Russians and people from the European Union. The articles and videos we had to watch before each class were interesting but sometimes a bit too strange. But in general, I always liked them and think they had a positive impact on me and my colleagues in class. After the class I often started discussions with my friends about what we learned in class to see how they would react on it. All my courses need to use this method of teaching because it makes the class more amusing and triggers me to learn about it even more. In Belgium and even on my exchange in Spain the student-professor relation is nothing compared to Russia.*

*I was not afraid to express myself even if it would contradict the opinion of the professor because I knew she could be able to understand me if I have the right arguments. It made me think more about how to express myself and address some sensible topics to a group. I learned a lot about interesting people in this sphere of sustainability and will from now on also look at the CSR report instead of only the financial one. I would have liked to have more guest speakers and learn about how they try to implement CSR into their businesses or have some interesting sustainable figures telling us their stories, like the persons from the TEDx videos. I am already trying to make a social impact and stimulate my friends to use a drinking bottle, for example, instead of buying plastic bottles. I hope my future career will allow me to stimulate society in a positive way and make people aware that living in a sustainable lifestyle not always means spending more time and effort.*

*Reflection # 6, HSE University student from Russia*

*The course on social responsibility of business was one of the best in terms of practical importance. It opened my eyes to the world around me, helped to look around more consciously, to see that everything is interconnected, that your specific actions affect society, to realize that by taking a small step with your actions, you are helping to change mindset of people to more sustainable. Now I am on my way to minimize my footprint. I started applying eco-friendly practices like using reusable bags and pouches instead of plastic bags; saving light and water. I start sorting the garbage for further processing, I am going to buy a reusable bottle for drinking tap water. One of the main things is that my example inspires other people to change their behavior and gradually, along the chain, we can improve this world together. On this Black Friday, I was not tempted to buy more things because of the big discounts. In general, I realized that Russia is not so bad with corporate social responsibility and environmental care. Through individual work and final work on AIM2Florish story, I learned many interesting companies, which previously did not know. And I realized that changes begin with us, with each of us.*

### *AIM2Flourish story for SDGs 8, 9, 10, 11 & 17: Equal opportunities for deaf taxi drivers*[5]

*Reflection # 7, HSE University student from Russia*

*The AIM2Flourish project as well as the whole course on social responsibility of business have inspired me and made me pay more attention to firms CSR and social projects. In my opinion, projects bringing the great value for society deserve and should be presented to publicity. AIM2Flourish platform collecting best practices of social responsibility inspires other companies to develop CSR and initiate projects in this field.*

*As for specifically our Yandex.Taxi project it was pleasant surprise for me to know that Russian company implemented such an important project for society and other companies. We were choosing between different possible companies basically those ones, where our team members are working. Since one of our team members works in the Yandex.Taxi and our professor advised us to take note to its project for deaf drivers, we had great opportunity to get deeper in acquaintance with the project and to analyze it from different perspectives.*

*Working in team, we have brought to the project our personal contribution in terms of diversity of views and experiences. Company was chosen without doubt because we all like this idea, but further we proposed questions for interview, discussed how particular part of AIM2Flourish report and class presentation should be described. Sharing of our views was significantly valuable for our common goal – to present the Yandex. Taxi project exhaustively and to get across the essence.*

*Initially when I found out about this project, I didn't realize the full significance of it, but AIM2Flourish platform and Ekaterina's mentoring helped to analyze it deeper and to make conclusions more conscious and global. The main lesson which I learned from the course and AIM2Flourish project is that sustainable business and social responsibility is not only under responsibility of the particular company, but also should be our area of interests in everyday life. This approach may increase the level of social responsibility overall and serve as a good base for sustainable development of business.*

*Reflection # 8, HSE University student from Russia*

*I did not realize that more than 13 million people with hearing problems live in Russia. This is a huge number. And I've never thought about how they live and how it is difficult to find a normal job for people with such problems. And I am proud that Russian company, Yandex Taxi help these people to have a normal job. I think that the project like this could help the deaf people integrate to our society. But I think that not only companies should help these people to live fully, but we also should do the similar things, trying to help them.*

*It was not easy to work in team, where everyone has a full-time job. It was real challenge to meet all together and prepare the questions to the interview. But when we had met with Daria and listened her story about this project I've understood it was worth it. I was impressed by the idea of this project and how Yandex thought of every detail. Drilling down to the project's details, I understood more and more that Yandex Taxi spend so much effort and energy, realizing this project. And I understood, that I also would like to have opportunity to realize similar project in the future.*

*Thanks to this project (I mean AIM2Flourish) and this CSR course I could develop the different helpful skills and extend my mindset. As I said, I did not think about the problem of deaf people, and I just could not imagine that the big companies such as Yandex could solve it. After this project, I will try to propose the similar cases to my working team to push them to think about it. I think that the more people will know about this and similar problems than more people with disabilities will live better and easier.*

*Reflection # 9, HSE University student from Russia*

*I have always been interested in social projects. Since studying at school I was a volunteer in various animal shelters, helped in the orphanage, collecting plastic from the streets of the city. But I always thought that it is very difficult to feel the contribution of one person, and to find a large number of like – minded people is not an easy task. But what can unite people, give them resources to implement social projects, and help to declare the work done around the world to inspire it to repeat? I realized – they are corporations! Only they have the most resources, talented people and potential to implement such initiatives.*

*I did not know exactly how companies can engage in social projects, who should be the initiator, how to evaluate the effectiveness. To understand this processes at all I began to study the course taught by Ekaterina Ivanova, and now I am sure that it was the most interesting and useful course for me, because it includes everything: an active teamwork, meetings with business representatives, a studying of social responsibility best practices of various companies, and understanding of the impact of business for the society. I fully realized that corporations are not only about how to increase sales and get the most profit. Corporations are open to social initiatives, they are ready to support them and talk about them.*

*I really enjoyed to work on the project with my team. We all work in different areas – marketing, consulting, audit, analytics. It was very interesting for me to get a different opinion from all the members of the team, which was based on the professional experience. Together we created questionnaire for interview, and if the marketing of our team tried to find out the image contribution of the project to the company, the consultants offered questions related to numbers and growth.*

*I really liked the course because it was very "alive". We discussed a lot in pairs, listened to speakers, read literature, watched videos, and interacted with business. Ekaterina Ivanova is a teacher who knows how to spark the interest in her work. I am sure that the amount of students, who understand that every business should implement a policy of social responsibility, has increased.*

*Thanks to this course, I realized that I would like to continue to engage in CSR in my professional activities and in my private life. Now I use less plastic, save paper, water and electricity. And this is still a small part of what I started to do and want to promote at the company I work for and in my family.*

## Evidence of impacts and lessons learnt

The students' reflections on AIM2Flourish showed that this experiential learning tool enabled them to question some of the tacit assumptions about the unsustainable anchors of their behavior, connect to their own values and believe in responsible and sustainable business as a new norm. The AIM2Flourish assignment was a transformational experience that increased students' self-awareness, prompted them to revisit and change some of their unsustainable habits and empowered them to influence others (friends, family and colleagues). Taking part in this project convinced even some of the most skeptical students that it is possible to do business with purpose, with respect to people, planet and profits at the same time. Russia as an emerging market is not an exception. That was one of the biggest discoveries for both the Russian and international students that jointly worked on this project. By connecting to real business leaders already contributing to achieving the SDGs, often without realizing it, students became inspired to follow their example. As a result of reflective practices, in-class discussions, and other introspective activities throughout the course, the students started a conscious turn to more sustainable behaviors in their daily lives. Stories of such changes in daily routines and aspirations on the choice of future employer and the role they intend to play in their professional life, which students shared in their reflective essays, show the impact of the learning they had at a deeper personal level. The experience of being a reflective story-teller as part of the AIM2Flourish project serves as a useful framework that helps students develop their sustainability mindset. The impacts of students' transformative experiences are visible in the three areas of *knowing* (e.g. deeper understanding of CSR and sustainable development, business models in Russia that are sustainable, the implementation of the SDGs, applying Appreciative Inquiry), *being* (e.g. self-awareness, habits, values, reviewing life purpose, intention to make a difference, social sensitivity) and *doing* (e.g. selecting future employers, influencing peers and communities, making professional choices, and taking leadership roles).

In terms of value that AIM2Flourish brings to the academic community, probably the most important is its openness to both signatories of the UN PRME and schools that are on their way of becoming part of this global initiative. AIM2Flourish is a free platform that is user-friendly for both professors and students. Thanks to an excellent support team, any problems with the platform are immediately addressed. Another great feature that AIM2Flourish provides is certificates to participating students and business leaders. Once the course is over, an informal meeting of the educator with the students to celebrate the publication of the story on the AIM2Flourish platform and distribute AIM2Flourish certificates could become an incredible moment to remember. Moreover, many of the students use the certificates, putting them on their CVs and public profiles on LinkedIn. Thanks to a multi-stage communication strategy, each time the AIM2Flourish stories are made public is an opportunity to share the victories of students in the competition for the 17 Flourish Prizes in a post on the corporate website and on the social media channels of the business school and businesses. Even within one business school, more than one professor could be involved in using AIM2Flourish as a learning tool; inspiring colleagues to use this tool proved to be a rewarding experience of sharing best practices. This program could also serve as evidence of business schools' contributions to achieving their third mission by addressing the real needs of society; AIM2Flourish offers a credible source of data for non-financial reporting, as well as for international accreditation and certification purposes with respect to the societal impact.

Among the most important lessons learned about using AIM2Flourish as a tool to shift students' mindset is that it works best when the educator takes the role of a learning facilitator and engages students on the individual level by facilitating a wide range of classroom discussions based on assigned course reading and video materials, interacting with guest speakers and organizing site visits (e.g. companies, museums, etc.). An educator's decision to let students work in a small group setting of up to three to four people would be beneficial for having positive group dynamics, avoiding conflicts and free-riding. AIM2Flourish is better suited for graduate students. For undergraduate students and non-native English speakers, it could become an extra workload for the educator to edit the narratives of the stories. AIM2Flourish is also an excellent gateway to get to know students better. After taking the course, many of the students ask for letters of recommendation, where the AIM2Flourish story could be integrated as a central element. With top students, such evidence helps them to get enrolled in highly competitive postgraduate programs in leading universities.

AIM2Flourish is already highly appreciated by educators around the world for helping their students to recognize how business is becoming instrumental in dealing with the challenges of the UN's SDGs, and what students can do to convert to better versions of themselves capable of contributing to a better world. The free platform with AIM2Flourish stories serves as a source of inspiration for thousands of educators and students around the world. As a next step, there is a potential to convert AIM2Flourish stories into teachable cases to be used by educators around the world. Also, a study on how the AIM2Flourish experience impacted participant lives five years from the launch of this initiative could be of great value to all of the involved stakeholders.

## Notes

1 As this chapter is being submitted for publication, two stories from Russia written by the HSE students have been awarded the 2021 Flourish Prizes for SDG 4: Developing Security Awareness Learning Journey for Everyone and SDG 17: Rescuers Finding Missing People with AI Technologies, see https://aim2flourish.com/2021-flourish-prizes.
2 AIM2Flourish, 2018; see https://aim2flourish.com/innovations/enabling-russias-digital-future.
3 AIM2Flourish, 2018; see https://aim2flourish.com/innovations/the-road-to-a-sustainable-future.
4 AIM2Flourish, 2018; see https://aim2flourish.com/innovations/changing-russian-recycling-culture-mission-impossible.
5 AIM2Flourish, 2018; see https://aim2flourish.com/innovations/equal-opportunities-for-deaf-taxi-drivers.

## References

AIM2Flourish. (2021). AIM2Flourish [online]. Available at: http://aim2flourish.com/.

Billett, S. (2009). Realising the educational worth of integrating work experiences in higher education. *Studies in Higher Education, 34*(7), 827–843. doi:10.1080/03075070802706561

Buchter, M. (2018). AIM2Flourish: An agent of world benefit. *International Journal of Appreciative Inquiry, 21*(3), 54–58.

Caffarella, R.S., & Barnett, B.G. (1994). Characteristics of adult learners and foundations of experiential learning. In L. Jackson & R.S. Caffarella (Eds.), *Experiential Learning: A New Approach* (pp. 29–42). San Francisco, CA: Jossey-Bass.

Cooperrider, D., Whitney, D.D., & Stavros, J.M. (2008). *The Appreciative Inquiry Handbook: For Leaders of Change*. San Francisco, CA: Berrett-Koehler Publishers.

Dean, B.A., Gibbons, B., & Perkiss, S. (2018). An experiential learning activity for integrating the United Nations Sustainable Development Goals into business education. *Social Business, 8*(4), 387–409. doi: 10.1362/204440818x15445231830058

Dewey, J. (1938). *Experience and Education*. New York: Collier Books.

Donaldson, T., & Walsh, J.P. (2015). Toward a theory of business. *Research in Organizational Behavior, 35*, 181–207. doi:10.1016/j.riob.2015.10.002

Elkington, J. (2020). *Green Swans: The Coming Boom in Regenerative Capitalism*. New York: Fast Company Press.

Fairfield, K. (2018). Educating for a sustainability mindset. *Journal of Management for Global Sustainability, 6*(1), 21–44.

Farias, G., & Waddock, S. (2018). Leaders as activists for a flourishing planet. *International Journal of Appreciative Inquiry, 20*(2), 77–90.

Freeman, R.E., Parmar, B.L., & Martin, K. (2020). *The Power of And: Responsible Business Without Trade-Offs*. New York: Columbia University Press.

Howard-Grenville, J., Davis, G.F., Dyllick, T., Miller, C.C., Thau, S., & Tsui, A.S. (2019). Sustainable development for a better world: Contributions of leadership, management, and organizations. *Academy of Management Discoveries, 5*(4), 355–366. doi:10.5465/amd.2019.0275

Kassel, K., & Rimanoczy, I. (Eds.). (2018). *Developing a Sustainability Mindset in Management Education*. New York: Routledge.

Kolb, D.A. (1984). *Experiential Learning: Experience as the Source of Learning and Development*. Englewood Cliffs, NJ: Prentice Hall.

Landrum, N.E., & Ohsowski, B. (2017). Content trends in sustainable business education: An analysis of introductory courses in the USA. *International Journal of Sustainability in Higher Education, 18*(3), 385–414.

Laszlo, C., & Brown, J.S. (2014). *Flourishing Enterprise: The New Spirit of Business*. Stanford, CA Stanford University Press.

Mackey, J., & Sisodia, R. (2013). *Conscious Capitalism: Liberating the Heroic Spirit of Business* Cambridge, MA: Harvard Business School Press.

Mezirow, J. (2009). Transformative learning theory. In J. Mezirow & E.W. Taylor (Eds.), *Transformative Learning in Practice: Insights from Community, Workplace, and Higher Education* (pp. 18–31). San-Francisco, CA: Jossey-Bass.

Mezirow, J., & Associates (Eds.). (1990). *Fostering Critical Reflection in Adulthood: A Guide to Transformative and Emancipatory Learning*. San Francisco, CA: Jossey-Bass.

Naim, A.I. (2018). Spiritual development programs, individual spirituality and sustainability mindset toward higher commitment to social and environmental impact. *International Journal of Business Studies*, *2*(3), 150–163.

Nonet, G., & Petrescu, M. (2018). AIM2Flourish: Students connecting with businesses doing good for our own good. In P.M. Flynn, T.K. Tan, & M. Gudić (Eds.), *Redefining Success: Integrating Sustainability into Management Education* (pp. 215–226). New York: Routledge.

Rimanoczy, I. (2017). *Big Bang Being: Developing the Sustainability Mindset for a New Planetary Imprint* (2nd ed.). Sheffield, UK: Greenleaf Publishing.

Rimanoczy, I. (2021). *The Sustainability Mindset Principles: A Guide to Develop a Mindset for a Better World*. New York: Routledge.

Sternad, D., Kennelly, J., & Bradley, F. (2016). *Digging Deeper. How Purpose-driven Enterprises Create Real Value*. Salt Mill, UK: Greenleaf Publishing.

Storey, M., Killian, S., & O'Regan, P. (2017). Responsible management education: Mapping the field in the context of the SDGs. *The International Journal of Management Education*, *15*(2), 93–103.

Vygotsky, L.S. (1978). *Mind in Society: The Development of Higher Psychological Processes*. Cambridge, MA: Harvard University Press.

Webster, K. (2017). *The Circular Economy: A Wealth of Flows*. Cowes, UK: Ellen MacArthur Foundation Publishing.

Wersun, A., Klatt, J., Azmat, F., Suri, H., Hauser, C., Bogie, J., Meaney, M., & Ivanov, N. (2020). *Blueprint for SDG Integration into Curriculum, Research and Partnerships*. Retrieved from New York: Principles of Responsible Management Education.

Wertsch, J.V. (1985). *Culture, Communication, and Cognition: Vygotskian Perspectives*. New York: Cambridge University Press.

# 10

# GRASPING THE UNSUSTAINABLE IN SUSTAINABLE ENTREPRENEURSHIP EDUCATION

*Jan Hermes, Elina Hanhimäki, Veera Hartikainen, Leon Hecht, Martin Friesen, Camille Jules, Sini Liimatainen, Emma le Maître, Hanna Małaszkiewicz and Catarina Neves*

## Introduction: sustainable entrepreneurship education at a Finnish university

At the University of Oulu, Finland, an annual series of six entrepreneurship courses is offered to any interested student enrolled at the University of Oulu, on any stage of studies and from any discipline. One of the offered courses is the *Entrepreneuring for Sustainability* course. It targets students who are keen to learn about developing a business idea that meets the "new normal," i.e. a business idea profoundly based on sustainability. Sustainability, here, follows the notions of simultaneous social, environmental and economic developments which do not compromise future states of society and nature (Brundtland, Khalid, Agnelli, Al-Athel & Chidzero, 1987).

The course aims to explore entrepreneurship as a form of creating socially, environmentally and economically responsible change in society. To this end, personal values are first explored and then translated into sustainable entrepreneurial business ideas by means of the UN Sustainable Development Goals and respective indicators (United Nations [UN], 2015). These entrepreneurial ideas are then publicly communicated to internal and external stakeholders of the University to create awareness of global and local sustainability challenges and to raise interest in participating in the realization of the ideas. Underlying those activities is the aim to develop and foster a sustainability mindset which is necessary to create both resistance towards the "old normal" and relentlessness to tackle complex and multi-layered challenges such as the creation of a sustainable world.

The course involves three lecturers from the University of Oulu who contribute to students' learning experience both with their professional and academic experience. One lecturer is an assistant professor with a research focus on responsible leadership and more than 20 years of experience in leading the human resources of a medium sized high-tech enterprise in Finland. Another lecturer is a doctoral student who specializes in scenario forecasts in business modeling. The third lecturer, who is the first author of this chapter, is an assistant professor with a research focus on business ethics and responsibility and about four years of work experience in project management in large-scale industrial construction projects.

DOI: 10.4324/9781003229735-12

Additionally, business experts are invited to speak as guest lecturers mainly about how to communicate entrepreneurial ideas to specific audiences.

In this chapter, first the course content, learning target and methods are introduced. Then, the learning portfolio as summative evaluation method is explicated and students' portfolio excerpts from the 2019 edition of the course are presented. The excerpts are structured according to the sustainability themes of systemic interconnectedness, more general contemplations on sustainability, awareness of self and others and confidence in the lastingness of a mindset shift. In the final section, reflections, critique and recommendations on how to nurture a sustainability mindset shift are provided.

## Course: Entrepreneuring for Sustainability

The *Entrepreneuring for Sustainability* course typically takes place during the autumn lecture period of Finnish universities and the active teaching phase lasts for about one month. Usually, during that month, eight half-day long in-class sessions with students are organized. To respond to potential health-threats arising from the COVID-pandemic, those sessions with a lower degree of in-class interaction are held online. Interaction-intensive sessions are organized as contact teaching with infection preventative measures and are supplemented with an on-line learning environment (Moodle) through which additional learning materials are provided and communication with the lecturers is possible outside class. Additionally, students are requested to familiarize themselves with lecture materials and perform follow-up course work between classes. The total workload for students is 135 hours (equaling five ECTS (European Credit Transfer and Accumulation System) credits), of which 36 hours are designated to in-class teaching, 20 hours for students' self-study of course literature, 80 hours for group work and 35 hours for writing an individual learning portfolio. Typically, the course attracts both domestic and international students of diverse backgrounds and with an equal gender distribution. This has led to rather heterogeneous classes of around 30 students.

The main mode of learning is collective learning, i.e. each participating student becomes a part of a student group soon after the beginning of the course in which the student remains until the end of the course. The course is rooted in a set of tenets that guide its planning and implementation: (1) reciprocity, i.e. a mutual balance of giving and receiving which means that everyone in class can learn from others and teach others and that there are no traditional learner and teacher roles, (2) experience, i.e. all activities are based on the experiences of participating students, lecturers and other experts, (3) personalization, i.e. every student (and lecturer) is appreciated for their individual background, and individual goals and expectations from the course, (4) experiment, i.e. in-class sessions are like laboratories in which there exists no right (nor wrong) answers, and (5) processuality, i.e. the course is regarded as a part of students' (and lecturers') life-long learning journey with no fixed stages of knowledge acquisition.

According to the University's syllabus for the course (Entrepreneuring for Sustainability [EFS], 2019), the following learning outcomes are intended:

> *Students develop skills for creative problem solving; students understand that entrepreneurial behavior can take place within many contexts (new ventures, associations, government agencies, and existing businesses); students identify their alternative roles, opportunities, and viewpoints regarding entrepreneurial choices they can make; students strengthen their skills of responsible*

> *business and are able to assess choices for business as promotor of social change based on the UN Sustainable Development Goals; students are able to define and assess alternative contexts for entrepreneurial action and to create and assess alternative business scenarios for their future; students are able to recognize and analyze business opportunities and social/customer problems and challenges; students are able to create and evaluate alternative solutions to the identified opportunities, problems, and challenges of responsible business; students are able to communicate effectively about their entrepreneurial ideas.*

To ensure the course outcomes, respective learning activities are grouped in three phases which describe a journey from the personal to the creative and to the public; it builds on the idea that the development of a sustainability mindset necessitates the engagement of head, heart and hands (Kassel, Rimanoczy & Mitchell, 2016; Hermes & Rimanoczy, 2018). The journey starts with students' exploring their heartfelt, personal values, continues with the translation of those values into an entrepreneurial endeavor that helps solve social, environmental and economic challenges and culminates in inspiring others for the endeavor. In the first phase, students are individually guided to explore their personal values as something that is close to their heart, functions like an inner intuitive compass, and gives direction in every action they take. Oftentimes, the values pertain to love, care, equality and other moral qualities. Awakening to one's own values provides the inertia for striving for a sustainable world. The second phase builds on the explication of personal values and aims at turning them into tools for the creation of sustainability. In heterogeneous (gender, discipline, cultural background) groups, students identify a sustainability challenge which they collectively feel dedicated to. The challenges are first concretized by means of the UN SDGs and respective indicators. Then, the students create solutions to tackle the challenges through entrepreneurial ideas. Iteratively, the development of future scenarios and the introduction of theoretical constructs from the field of business ethics, CSR and sustainability are supposed to help molding sustainable entrepreneurial ideas and, at the same time, students' sustainable mindset. The third and final phase is dedicated to going public with the entrepreneurial idea. Students learn how to articulate their solutions to sustainability challenges in audio-visual forms so as to attract attention of both potential participant/ customer and investor audiences.

The course is devised to incorporate the elements of thinking, being and doing, which together comprise the ingredients for the development of a sustainability mindset (Kassel et al., 2016), as shown in Table 10.1. While certain sessions have a narrower focus on e.g. only one element, other sessions more holistically include several elements.

Some of the eight in-class sessions end with a reflective question (see Table 10.1). The questions wrap up large parts of the preceding session and are intended to support students' reflection process. Taken together, the course reflections are compiled by students to create an essay ("learning portfolio"), which is used as a summative evaluation method. The learning portfolio requires the reporting of all relevant in-class content including evidence and integration of all course material. More importantly, however, is its emphasis on students' learning process; they are urged to articulate (textually, visually, or else) what they experienced in the course, to bring forth their own thinking and its development, the emotions and feelings triggered during the course and critically reflect on both in relation to the course activities.

**TABLE 10.1** Course session details

| *In-class session #* | *Topic* | *Activity* | *Sustainability mindset dimension* | *Reflective question* |
|---|---|---|---|---|
| 1 | Personal values and vision | Individually exploring personal heartfelt values | being | What will I see, have and be when I have achieved my personal goals? What values guide me in this? |
| 2 | Mindset and UN SDGs | Understanding the responsible mindset and the UN SDGs bases for its realization | thinking | What is an irresponsible mindset? What is at the core of meaningful business? |
| 3 | Idea development workshop I | Identifying and prioritizing stakeholders of a sustainable entrepreneurial endeavor | thinking, doing | What makes a stakeholder less important? |
| 4 | Scenario development workshop I | Developing alternative hypothetical futures of the entrepreneurial endeavor | thinking, doing | n/a |
| 5 | Scenario development workshop II | Identifying ideas, drivers, trends and potential events which together cause changes in the environment | thinking, doing | What future scenarios can you envision for your sustainable entrepreneurial endeavor? |
| 6 | Idea development workshop II | Understanding the economic, social and environmental realms of sustainable endeavors | being, thinking, doing | What makes the parallel achievement of economic, social and environmental realms challenging? |
| 7 | Personal and corporate branding workshop | Developing personal and corporate branding strategies | thinking, doing | n/a |
| 8 | Idea pitching | Presentation of groups' entrepreneurial endeavor | doing | n/a |

## Learning portfolio: Profound reflections as summative evaluation method

The given reflection questions serve as basis for students' compilation of their individual learning portfolio which is used for the evaluation of students' performance. Most of the questions are posed in an anti-phenomenal manner, i.e. they ask the student to approach a phenomenon from its opposite. For example large parts of session 2 were dedicated to understanding what a responsible mindset is. In order to support students to embrace the concepts of a responsible mindset, they are provocatively challenged to contemplate on the contrary, i.e. an irresponsible mindset.

Thirty-two students participated in the 2019 edition of the *Entrepreneuring for Sustainability* course and created learning portfolios that contained both textual and visual elements, such as photographs, self-drawn pictures and maps. In the following sections we present learning portfolio excerpts that portray instances during the course that triggered deeper thought processes with nine participating students. None of the students were native English speakers; yet their portfolio excerpts are left largely unedited to convey their authentic voices. These instances are categorized into four main themes: systemic interconnectedness, more general contemplations on sustainability, awareness of self and others, and confidence in the lastingness of a mindset shift.

## *Systemic interconnectedness*

Several in-class exercises as well as students' reflection work outside class suggest that they have realized the interconnectedness of the socio-material world we live in, as Camille writes:

> *I think that an irresponsible mindset is also characterized by the absence of change. It chooses having and being rather than doing and becoming, it focuses on past patterns and on self-beliefs and is not able of awareness so it cannot act according to a new context and will always repeat the same actions making wrong choices in its environment but always good choices for its self-interest and as it never questions the world, it remains in its own selfish loop hurting the evolving world around it. It led me to think about Giovanni Papini's sentence from Un Uomo Finito: "All the national, social, moral questions are, at the end, nothing else than soul questions, spiritual questions. Changing the inside, we change the outside restoring the soul, we restore the world." At the light of the concept of mindset, I interpreted this sentence as meaning that the acts we do according to our mindset have influence on the world around us, a responsible mindset will change the world in a positive way and an irresponsible mindset will change the world in a negative way trying to shape it for its own interest (as Papini did …).*

Likewise, Martin describes a process of awakening to the complexity of socio-economic systems and the necessary awareness of these interconnected systems as well as the role individuals and their decision play in the systems' development:

> *The lecture was started with a task in which we should deal with who and what made our coffee possible in the morning. It became clear to me that behind a seemingly simple decision like drinking a cup of coffee there is a greater complexity. Many different factors and people are involved in ensuring that my café arrives at me as it is. It became clear that this complexity and dynamic can be found in real life everywhere in the context of decisions. On the one hand, for my morning coffee I need a farmer who grows and harvests the beans, and on the other hand, my consumption affects him and his decisions. This can be also seen throughout the supply chain. So, to make decisions in a responsible framework, it means being aware of the complexity and knowing or questioning the impact of your decision. The other way around, you are acting irresponsibly by not questioning the consequences of some decisions. This also applies to small seemingly trivial decisions, since these also cause effects.*

Also, Veera writes about her understanding of the effects of individual decisions on the larger system we live in as well as how that system is more encompassing than initially assumed. She describes how a lack of this understanding leads to paralysis and a possible corroboration of the unsustainable status quo:

> *Another thing that I hadn't been thinking about before was all of the people my consumption may have an effect on. I hope everyone could live as happily as we here in Finland and that our consumption would affect as little as it could other people's lives but that is just not the case. In my decisions I have straight affect to the nature and ecosystem, to the economy of the world, maybe to my friends and co-workers and sometimes even to undeveloped countries and their people. This is something everyone should be having in their mind when they are thinking between different options and which to choose. The awareness and the responsibility of all of us should be taught to us at elementary school so everyone would learn to consume responsibly. Everyone has the duty to make things better in their own way and the way they can. Otherwise, we will be living in a cycle that never ends and we can't change the situations that are around us. I'm very worried … The next lecture was about creating more our business idea. We started to have some kind of picture in our heads about our business and what will it do, and, in this lecture, we got a bit further with the thinking. These steps that we took in this lecture were something I couldn't have come up without guidance. I learned a lot about what kind of steps we need to consider when we are planning to start a new business. Steps that I didn't know even existed. First, we talked about identifying our stakeholders. We used different categories to identify how "important" the different stakeholders are for us and in which way they will affect to our business.*

### *General sustainability contemplations*

For some students, the course and specifically the learning portfolio compilation in specific provided an opportunity to think about sustainability more generally. Martin, for example, describes how he thought more deeply about the constructs and phenomena discussed in class:

> *By taking a closer look at Ethics-as-practice, I recognized the irreciprocal, infinite responsibility that is anchored in it, which means that the café farmer is responsible for me and I also have a responsibility for the farmer. Furthermore, questioning this responsibility is an ongoing process (Clegg et al. 2007). It would be irresponsible not to assume any dynamism here and to assume fixed relationships that do not change. This insight must also be considered in the further course of a company and in my opinion, it is not sufficient to merely consider this during the development of a business idea. You should not rest on your laurels and go one step further … I think it was very successful to organize my thoughts through the Learning Portfolio. I often had the impression that I had dealt with core topics in more detail through this portfolio and had critically questioned and reconsidered several statements and models. For me personally, the course formed a very stable basis on the way to becoming self-employed and took away many questions. In combination with the scientific papers, a lot of information was available and, if a topic was unclear, this literature was a good point of access.*

Veera, too, mentions how e.g. the personal value exploration affected her thoughts outside class:

> *The first question was what kind of values we have. This was absolutely the first time I started thinking about my values and I kind of got lost in my thoughts because I had no idea what values are meaningful for me and have I been making my decisions based on what. We discussed that the values can be found when you meet some challenges in life or work etc. Also, the moments when you meet moral dilemma and don't agree with some kind of thinking. I had never been before thinking that there are values in different levels like us individuals but also organizations, countries and world have own kind of values. I wanted to go bit further thinking about my values and later on the week I realized that probably the most important values for me are my dear man and dog, healthy lifestyle and getting good education leading to the work I want to do.*

Catarina contemplates on a more general level how responsibility is infinite and ubiquitous as well as what consequences that line of thought would have in today's world:

> *For me, Levinas says that we have infinite responsibility in our actions, in our decisions and in the way we conduct our life. Whenever we decide to do something, or even decide not to do anything, that decision is our responsibility. Taking the example given in the class, if someone I know fails an exam, the responsibility of failing the exam belongs to the person that failed and only to her. But the way I react towards the situation of this person, is my responsibility. In this situation, I believe that this principle defends that if I have the power to act in such situation, if somehow I can influence the circumstances of that person, and this can be made by all kinds of action, from a kind word to an advice or a suggestion of a better method to study, I am responsible for the way I choose to act or I am responsible for choosing not to do anything. For me Levinas was trying to say that we are responsible for everything we can influence. Our decisions and actions affect our surroundings, no matter how little that effect is, and that influence is not only reflected at the moment a decision is made, which is exactly why this principle transcends time and the notion of responsibility is infinite. Even though I'm conscious that this principle should not inevitably dictate our course of action because we would end up in an infinite spiral of indecision and possible scenarios of how broad our influence might be, I believe that, when having a responsible mindset we should use the principle of infinite responsibility to keep in mind that what we decide now doesn't affect only us and that the present has influence in the future.*

## *Self- and other-awareness*

The course proved to be awareness-creating in two ways to some students; on the one hand, it triggered thoughts about themselves and, on the other hand, about others as well as how the self and the other are related. Emma, for example, describes how increased self-awareness helps understanding and being compassionate toward the other, too:

> *As we learned with the teacher, being aware of our values helps us push forward and move on in our professional and personal life. I also learned that finding our values influences our choices. I totally agree, for example I'm curious, I love discovering other cultures, behaviors, different ways of learning etc., that's why I decided to experience Erasmus. And thanks to this experience I've learned about myself and I have another point of view about school and classes compared to France. I believe we have to see what's happening outside to understand the world we are surrounded by. I also believe it has an impact on our future and how we think about businesses.*

Hanna writes about a process of self-confrontation in which she became courageous enough to introspectively uncover the values that are at the bottom of much of her being and acting:

> *In everyday life we don't really think what concrete stimuli is under our behavior. We clearly can decide what is acceptable for us and how we perceive certain situations, but on my personal example I can say that I have never asked myself "what are my core values?" I often heard this question during Miss Universe contests and usually disregard contestants' efforts to answer that but wasn't brave enough to ask it myself. Now I had the opportunity and evaluating this brought me satisfaction ... I'm a big dreamer, but never brave enough to say my plans out loud. Creating my vision map was quite scary and liberating experience for me and brought me big satisfaction. I realize that it can be done and set these "dreams" in my mind. I find it as very useful tool for self-motivation.*

On a more critical note, Leon writes how he was able to reveal something previously unknown to him, but urges for a more elaborate way to do so:

> *I do not think that a vision map suffices for displaying every single hope, wish and desire. Reducing all expectations for the future to this level can lead to dismissal of idea diversity. I suppose most vision maps in the course are quite similar, since health, family and money are commonly liked aspects. I would have wished for a tool allowing portrayal of more perspectives, making greater diversification possible. The process hiding behind this depiction however I found very interesting. Evaluating what really matters in life, especially concerning the future, turned out to be more difficult than expected. While in the beginning, only the greater clusters were visible to me, further questioning revealed quite a lot about myself, that I was not conscious of before.*

### Confidence in lasting mindset shift

Finally, some students seem to have gained confidence in the sustainability of a freshly created sustainability mindset. Elina, for example, is optimistic to maintain a sustainability mindset in both her private and professional life:

> *The journey we have gone through in this course has been a lot more rewarding than what I imagined in the beginning and I feel like I have learned tremendously during these weeks and long afternoons sitting in the classes. I will hold onto the responsible mindset we discussed in the lectures and consider my own behavior as a consumer and as a possible future business owner.*

Similarly, Sini writes about her determination to think sustainably as consumer and entrepreneur:

> *Before this course I was already really interested about the connection between entrepreneurship and sustainability but sustainability itself was the main point why I chose this course. I realized that sustainability is more than I previously had thought. Before, I just thought it was all about the environment but during this course I realized that it had multiple other dimensions for example the social side. I found all the lessons to be really interesting. I really liked that we had time to go through all the SDGs because I wasn't familiar with them previously. I also really enjoyed working with our group on the business idea and getting more experienced with*

*talking in front of the class. After this course I want to continue thinking in a sustainable way when consuming and making decisions. This course will be helpful in the future if I want to start my own business.*

Veera describes how she already during the course started enacting a sustainability mindset in her every-day decisions:

*The lecture started by thinking about our coffee drinking and what are all of the things that are affecting to it and to what effects. This was a great way to start thinking about sustainable way and set up our minds to the right place regarding that. I started using some things from this conversation straight away at home when I buy something, I try to think that is it's sustainable decision to do. Coffee effects to lots of things and I really want to avoid irresponsible mindset and just act without thinking about the effects of my acts. I have been doing it in a small scale at home, trying to cut out as much of energy and water use and avoid waste.*

## Reflections, critique and recommendations

The learning portfolio excerpts portray how students profoundly reconsider hitherto unquestioned assumptions about (un)sustainable ways of doing business and consuming and, more generally, their ways of leading life. The contemplations pertain to realizations of systemic interconnections and other responsibility matters, the creation of awareness of self and others and the sustainability of the sustainability mindset.

What is common to most of these contemplations is the way they were stimulated, i.e. by means of asking students to explore the anti-phenomena. The learning portfolio excerpts show, e.g. how explicating the characteristics of irresponsible and unsustainable business conduct supports grasping the opposite, i.e. responsible and sustainable business conduct. Hereby, of help is not only the rational clarification of a phenomenon's stark contrast (following the idea that one can only know what "warm" is if one knows what "cold" is), but also the emotional power of dwelling into the ruthlessness, cold-heartedness and profound injustice that unsustainable business can produce. Students' learning portfolios also suggest that a strong connection of the course content to their own person, i.e. their understanding and feeling, is key to triggering self-awareness and self-confrontation processes. Whether or not such processes are fruitful surely depends, among other things, on the students' willingness to open their eyes to taken-for-granted core assumptions about life which oftentimes constitute the basis for their own being and acting. Moreover, the learning portfolios are artifacts of students' self-confidence; not only is there a remarkable amount of knowledge of business ethics and responsibility/ sustainability (which is not part of the syllabus) with some students, but also the level of applying and practicing responsible/ sustainable principles is high. Taken as a whole, students' learning portfolio reflections show a promising mindset shift-in-process.

The sustainability, or lastingness, of a sustainability mindset in the making is yet to be proven. Much like a young plant, also a new mindset requires continuous nurture and care. Resisting the existing traditional pressures by coworkers, customers, investors and others surely is not a cinch. Another point that deserves critical attention is that of the cultural setting in which the course takes place. It is sometimes challenging for Finnish people to profoundly disclose something as dear to their heart as their personal values. The same applies, at least to some degree, to international students who have been exposed to the Finnish culture for a while. Hence, interpreting these kinds of learning portfolio needs to be done in a culturally sensitive manner. The learning portfolio excerpts used in this chapter can be seen as proof that the course activities, particularly the introspective value exploration in the beginning, can move students to openly confront themselves with something ultimately intimate and potentially status quo challenging, i.e. their personal values.

Lecturers of sustainable entrepreneurship aiming to create a mindset shifting experience are recommended to design their teaching in a holistic way. In addition to the oftentimes dominant element of thinking, the doing and being aspects need to be emphasized to speak to students as *entire* human beings. A mindset shift seldom happens through a rational conclusion only but requires its practice and enactment as well as respective emotional work. It is particularly the being or emotional element which seems odd or too personal to some students who are used to and expect rational and fact-based transmittance of knowledge. At the same time, the being element is often one which leaves the most lasting and sustainable impression on students. In doing so, an anti-phenomenal learning method as a form of delving into what is *not* sustainable can support the grasping of the sustainable and help pave the path toward a sustainability mindset. In other words, reaching sustainability can require an excursion into the unsustainable.

## References

Brundtland, G.H., Khalid, M., Agnelli, S., Al-Athel, S., & Chidzero, B.J.N.Y. (1987). *Our Common Future: Report of the World Commission on Environment and Development*. Oxford University Press.

EFS (2019). Entrepreneuring for sustainability course. *Syllabus*. Retrieved 15 July 2020, University of Oulu. https://www.oulu.fi/kauppakorkeakoulu/node/46251

Hermes, J., & Rimanoczy, I. (2018). Deep learning for a sustainability mindset. *The International Journal of Management Education, 16*(3), 460–467.

Kassel, K., Rimanoczy, I., & Mitchell, S.F. (2016). The sustainable mindset: Connecting being, thinking, and doing in management education. In *Academy of Management Proceedings, 2016*(1), 16659. Briarcliff Manor, NY: Academy of Management.

UN (2015). Sustainable development goal adoption, *UN resolution 70/1*. Retrieved 15 July 2020, United Nations General Assembly. https://sustainabledevelopment.un.org/post2015/transformingourworld

# 11

# STUDENTS AS CHANGE MAKERS TO ACHIEVE THE SUSTAINABLE DEVELOPMENT GOALS

*Morgane M. C. Fritz*

## Introduction

The Sustainable Development Goals (SDGs) set by the United Nations (UN) in 2015 (UN, 2015) have been increasingly used for research, teaching and practice, to the extent that they are now becoming a framework or guidance for some researchers, teachers and firms. Sustainability reports, for instance, are a way of exemplifying the use of SDGs by firms and educational institutions. However, SDGs are mainly used to highlight what organizations or researchers do in relation to sustainability and employ a descriptive approach, since to date there are no indicators to measure an organization's contribution to the goals. Consequently, it is rather rare to find analysis of how to put SDGs into practice. The aim of this chapter is to present a more concrete implementation of SDGs as explored with a group of students from the Excelia Business School (EBS) in France, through an exercise the author developed and presents here.

The exercise has its roots in the sustainability mindset (SM) learnings from the SM fast track course developed by Isabel Rimanoczy and which the author followed in 2020. Isabel Rimanoczy is the global expert on how to build an SM and she developed a framework to explain and encourage the development of this mindset (Rimanoczy, 2017, 2020). This chapter focuses on the "doing" side of the framework, whereby students, after following a lecture on sustainability and ethics (given by the author of this chapter), were asked to choose one SDG. Their task was then to undertake an action related to that SDG during the time period of the course and to write an essay to justify their choice of SDG, explain the action they took and highlight their feelings and any ethical values associated with the exercise.

This chapter specifies the content of the lecture given prior to the essay exercise, presents the detailed task given to the students and shares some of the most relevant contributions from students who gave their consent for the author to do so prior to publication. The findings show that the students were able to accomplish actions that they considered small but that provoked a change in their mindset and sometimes even in their family's or friends' mindset. The findings also show that acting in relation to the SDGs opened the students' eyes to the impact of individual actions in a global context through observations they made about their

DOI: 10.4324/9781003229735-13

environment or their own being. At the end of the chapter, the author provides feedback and recommendations for those researchers and teachers who would like to replicate the exercise. The outcome of this exercise is a panel of messages full of hope that transformational practices can be accomplished through individual change.

## Context in which the essays were produced

The essay was one of the assignments given to a group of 34 students in 2020 during an elective called Business Ethics and Corporate Social Responsibility at EBS in France. This elective took place during the second year of the Grande Ecole Programme (Master's in Management). Students had already taken classes on theories related to business ethics and corporate social responsibility (CSR) during the first year of their Master's programme. The lecture started with theoretical inputs on the concepts of business ethics, CSR and the UN SDGs, as well as examples of what companies can do well or get wrong. The teaching style adopted was to focus on developing the students' critical thinking by regularly asking them their point of view and giving them time to think about the different concepts and examples seen in class. Of the 34 students, 17 agreed that extracts related to their reflective essays could be shared anonymously.

## Learning goals that prompted the selection of the activity

At EBS, teachers have to take five learning goals into account in the design and content of their courses and to make this clear in the syllabus. These five learning goals are as follows: (1) curriculum and teamwork; (2) international perspective; (3) oral and written communication; (4) ethics considerations and corporate social responsibility; and (5) research. It is worth noting that ethics is already part of the learning goals and thus any lecture can include an ethical perspective. All five learning goals were selected for the Business Ethics and CSR elective.

Beyond this learning goal system, each teacher at EBS can define more specific learning objectives. For this elective, the emphasis was set on: (a) raising students' awareness of business ethics and CSR issues; (b) developing ethical leadership; (c) understanding and contributing to ethical and responsible decisions in businesses; (d) identifying, analyzing and solving ethical and CSR issues; (e) experiencing conflict between personal values and organizational goals; and (f) developing an ethical and sustainability-oriented organizational culture. These specific learning objectives are intended to prepare students for their working life and anticipate the day they may be part of an organization. The reflective essay the students were asked to prepare focused on how the specific learning objectives (c), (d) and (e) applied to the students' own life experience, since most of them had not yet worked in a company for any length of time.

## Details of the activity and the assignment that prompted the essay

The course was composed of different individual and group activities. The essay was one of the individual assignments. The assignments were given to the students at the beginning, so that they were aware of their workload, could ask questions before the deadline and start whenever they wanted.

**TABLE 11.1** Individual assignment given to Master 2-level students

**Individual essay assignment**

Choose 1 SDG and take 1 action that contributes to it positively between 12 and 17 of March.

Questions:

1) Explain what SDG you chose and why
2) Describe the action you took and why
3) Explain what outcomes you saw from your action, and
4) What you learned from doing that
5) Based on the action you took, which is linked to a specific SDG, can you think about any positive or negative consequences on another SDG? How is the action you took for 1 SDG related to one or several other SDGs?

Write 2 pages.

Deadline: 17 March. Upload you work on MyLMS before 17 March 5pm.

Volunteers to present their work on 18 March.

Note: what is important here is not the number of relations you find, but really that you critically think about the existing or potentially existing relations, the quality of your reflection and argumentation. A good reflection and argumentation considers for instance the pros and the conts, the what/why/who/where questions, and is illustrated by examples.

Good luck!

Best regards,

Dr. Morgane Fritz

This chapter is entitled "Students as change makers to achieve the Sustainable Development Goals" because I did not want to teach them what is right or wrong in terms of doing business ethically and supporting the SDGs. I wanted the students to discover this for themselves; "learning by doing" as we say, and in order to "foster them to shape the world they want to live in," in the words of Isabel Rimanoczy (2016, Abstract). The objectives were to make the students active rather than passive and to enable them to realize that they can all have a role to play in promoting a more sustainable world. Despite the short amount of time, they had (five days), I required them to take an action related to an SDG of their choice and to write a two-page essay about this action, supported by the questions shown in the assignment below (see Table 11.1).

This assignment was inspired by an experience Isabel Rimanoczy shared with the PRME community of requiring students in Morocco to develop a project in a few days and the strong impact this experience had on their sustainability mindset. Therefore, I thought it should be possible to have the students I was teaching develop and experience actions that were impactful for them and, eventually, the community of which they are a part (e.g. family and friends).

## Comments about the students and challenges encountered

Throughout the course, the students were really active, involved, and were asking and answering questions, probably because they were on an elective. When asking them what brought them to the class, several answered that it was because they found their first-year introductory lecture that I gave interesting and wanted to learn more about CSR and business ethics.

Some challenges arose when students started to work on their reflective essay. Several of them had experience of working as a volunteer for a humanitarian project, which is a requirement of their *Humacité* programme at EBS.[1] Students are required to take part in a volunteering experience, either in France or abroad, and most of them choose missions related to children's education in developing countries. Since this voluntary work was recent, several students asked if they could write their essay based on that experience. Although this might also have been relevant, I decided not to accept these requests because at the time they did their voluntary mission, they were not in the process of contributing to an SDG and they might not have recalled the feelings they had when undertaking specific activities within their mission.

Once it had been made clear that all the students had to take an action related to one SDG and report on their learnings, observations and feelings related to that action, they were able to carry out the assignment. The assignment started in class (about 30 minutes) and continued as homework.

When I received the essays and read them to grade the students, I realized that several of them had not fully understood the exercise, as they explained one of the SDGs instead of taking an action related to it. These students were, however, still able to learn something during their research. For instance, one student was surprised that many actions have been undertaken for SDG 5 ("Achieve gender equality and empower all women and girls") but she could still observe a number of disparities and inequalities between the genders, and she did not understand why. Several students did not follow the assignment, despite the clarification. They reported, for instance, on past experiences or on their lifestyles. Several of them were nonetheless able to produce excellent essays and reflect on the effects the action had had on them, on the environment and on their community (e.g. collecting plastic from a beach).

I have some assumptions about why several of the students did not follow the assignment. During the class, some of them had looked at me as if I were asking them to climb Everest ... They were probably not used to being asked to do something by themselves in class, which is a typical issue in the French education system. As I am myself French, I can attest to this: students typically spend hours of their day listening to a teacher and taking notes. One can notice the huge difference with students coming from abroad, especially from Northern or Eastern Europe: they are much more used to doing assignments on their own and reflecting on what they have learned. Another assumption is that some of the students genuinely did not understand the exercise because of language issues, which, based on my teaching experience, is a recurrent issue among students in French classrooms.

## Analyzing students' reflective essays in terms of content and impact

The analysis of the students' essays revealed insightful contributions and outcomes for them and the SM principles. Students that successfully completed the assignment had various answers and experiences. Most of them chose simple actions, such as doing the food shopping and paying attention to buying only local products or not buying from supermarkets. Some tried to buy food without any plastic packaging and found it very challenging. Others were not yet fully separating their household waste and decided to pay attention to that, and were surprised by the amount of plastic waste they were generating and the difficulties in sometimes finding out whether waste is recyclable. Other students focused on mobility and decided not to use their car or any public transport or other mode of transport that was polluting. These students ended up walking to the school and found it relaxing, or even felt good that they had not contributed to polluting the air.

What can be highlighted from these essays is as follows: (1) students' motivations for choosing one SDG rather than another; (2) the various actions taken; and (3) the various outcomes for the students' emotions, their environment, or their motivations for going further. The following are excerpts that illustrate these three points. These excerpts are shared here in the form in which the students wrote them. For many of them, English is not their mother tongue, which explains some of the grammar and spelling mistakes.

### Examples of students' motivations for choosing one SDG rather than another

- SDG 2 "Zero hunger": *"malnutrition should not exist"*
- SDG 3 "Good health and well-being": *"I have chosen that SDG because for me, health is one of the most relevant goal. Effectively, it is directly related to human dignity and essential needs: without health humans are nothing"*
- SDG 11 "Make cities inclusive, safe, resilient and sustainable": *"Urgent action is needed to reverse the current situation, which sees the vast majority of urban residents breathing poor-quality air and having limited access to transport and open public spaces. With the areas occupied by cities growing faster than their populations, there are profound repercussions for sustainability. I know in near future I will probably spend a part of my life in a big city and so far there is a hundred of pros living in those cities but there is also a number of cons: pollution, lack of transport, property prices etc."*
- SDG 12 "Sustainable consumption and production": *"This particular objective interest to me because I find it accessible to everyone … The current consumption problems are obvious. Indeed, in the world today, we throw away a third of the food produced, which is equivalent to 250 cubic kilometers. We also see that the planet's resources for a year are exhausted after barely 8 months. If people learned to consume more responsibly, we could better equalize resources: for example, limiting hunger in poor countries, but also reduce pollution. On the other hand, money can also be saved, indeed, if all humans started using high-efficiency light bulbs we would save 120 billion dollars every year."*

### The various actions taken by students

- Donations (food, blood)
- Composting vegetables
- Volunteering for charities
- Waste collection.

### The various outcomes for students' sustainability mindset or feelings

- **Personal satisfaction**

> *"we are also glad to see the smiles on the faces of the patients, we showed them that everybody is equal and every life deserves a nice way of treatment. The happy faces and the mental relief that we witnessed from those who we've talked with is the most precious memory of all time."*

*"at the end of this day I was really happy to see how we did well, almost 64 KG of wastes."*

*"I didn't do it before and I regret that but now, I am glad to give blood as regularly as I can. I do that action to help people in need of blood. I know it is a quite small action to help for a healthier society, but it matters for me, and I think it is yet something positive and helpful at my scale."*

- **It is easy to contribute to sustainability**

  *"I didn't think it was that easy to make and self-regulating compost. It's a way of recovering your waste without being time-consuming. Compost is 'self-regulating.' The only constraint is that you have to have space outside to make it."*

- **Questioning**

  *"I asked myself a lot of questions. Why are there still people in famine or malnutrition? I'm so far from realizing the problem, going shopping in a supermarket that answers very little to an ethical question, that wastes … This gesture makes me want to do more and more often."*

- **Relativism**

  *"By doing this, you realize once again how bad the world is and that not everyone has had the luck and comfort that I have had. I've also seen people refuse to give, even a packet of pasta and it's not easy to see. Without forgetting that we should not judge but it is sometimes difficult to see citizens turn a blind eye to that."*

- **Disgust**

  *"Giving for nothing in return is not easy for many … The result is ultimately disgust, disgust for my lack of investment and that of others but even more compared to those who close their eyes to avoid 'feel guilty' and not see reality in the face."*

- **Changing habits**

  *"The selective sorting of household waste imposed by the urban community of Lille has made us aware that everyone is responsible for their actions. We have therefore changed our way of consuming in order to reduce our waste."*

- **Willingness to do more**

  *"This is the first step but I believe also, we need to unit to get more impressive when we are taking actions. If I was by myself during cleaning the beach and the water, I'll be disappointed about my work, about the impact of my action, but because we were few peoples the beach seems really clean, and it's a great success, not only for our environment, but for our mindset."*

Overall, by looking at the students' motivations for choosing an SDG, the action they took and the impact it had on them, several points can be highlighted from my perspective. First, it is highly relevant to develop exercises where students become actors, become change makers, and what we see from the quotations is that, by themselves, the students started changing their own mindset first, not the mindset or actions of others. Although critical thinking is difficult to stimulate in class, the students showed that they can be critical of themselves and the overall production and consumption system in which we live. This might be an indication that some students feel more comfortable being critical in a written form or in a one-to-one interaction, rather than in a classroom. In class, several of the students did not feel at ease reporting on their feelings; they preferred to report on facts. However, it was easier to find indications of their feelings in the essays (e.g. disgust, satisfaction or happiness). Social and emotional learning, such as through this simple exercise, are essential to achieving the SDGs so that individuals can identify conflicting goals and values between what they should do, what they actually do and what society expects them to do (which refers to Festinger's dissonance theory from 1957).

In order to achieve the SDGs, Asah and Singh (2020) consider social and emotional learning (SEL) an essential step in bringing individuals closer to the meaning of the SDGs and encouraging their engagement with them. Several contributions support this argument in practice. For example, Shrivastava (2010) argues that sustainability needs to be taught in a way that mixes both intellectual learning but also physical and emotional learning in the context of a sustainable management course. Being able to stimulate students' feelings in relation to sustainability is a key to supporting change making alongside intellectual and behavioral forms of learning. Of the different forms, emotional learning "is the most powerful one, as it makes the awareness more profound and grounds it, is more lasting even if we 'forget,'' it comes back" (personal communication with Isabel Rimanoczy). Ives et al. (2020, p. 215) call these feelings the "inner worlds of individuals," which, they argue, research and practice have neglected for too long, even though "these have the potential to fundamentally shape human behaviour and possibly even the functioning of social systems."

## Follow-up contact with the students

Following the end of this class, the International Students' Sustainable Development Week (SSDW) was taking place, virtually, because of the COVID-19 pandemic, from 6 until 12 April 2020. Some of the students' essays were of a high quality, so I proposed organizing an event that would allow those students who agreed to do so to share their essays. A small group of two students quickly took up this idea and managed to persuade the others to produce short quizzes and facecams on Instagram to raise awareness in the school of their SDG-related action. This group was encouraged by one of the students, who was the President of the EBS Student Association for Sustainability (called "Oxygen"), and by a request from the EBS Communications Department to comment on our activities via EBS social media and the REFEDD organization, which represents all French students' associations for sustainability.[2] With my support and the help of the Communications Department at EBS, five students contributed to the SSDW by extracting information and experiences from their colleagues' essays that they found relevant to disseminate (with those students' agreement). I was surprised that these students were willing to spend quite some hours on this communication event to disseminate their work and it encouraged me to do an introductory facecam as well (for the first time in my life!). The students were really grateful for being given this opportunity to share their work.

Some months after, in September 2020, two students came back to me to ask if I would like to supervise their Master's dissertation because they really liked the course. What great feedback, to see that six months later they still remembered what we did together and were willing to dig further into business ethics!

## Conclusions and recommendations for replicating the activity

To conclude, having students write an essay is a well-known assignment that can be used to encourage the students' sustainability mindset, combined here with the "doing" part of the SM. Requiring students to take an action and reflect on it within the SDG framework is an interesting way to have them implement the Goals and analyze them in a context with which they are familiar (their own life), rather than a setting that they do not necessarily know well yet: a firm. Whatever the context, students experience, and will experience further when they are part of an organization, the need to make decisions and to reflect on the impact their actions and decisions may have on others and the environment. I believe this exercise is relevant for all students studying management, since they will be increasingly required to assess the sustainability impacts of companies' activities. If they have experienced this at an individual level, they may be able, willing, or may even take the initiative, to conduct such analysis within the business context. Still, it remains the case that this was the first time I had designed and given such an assignment and there are opportunities to further improve it, as follows.

- Be very clear concerning the instructions: either accept that students can write about past experiences or reject this possibility and state it clearly in the instructions. Essays on past experiences can also be very rich, but those on actions the students have to take clearly show them that they can be change actors and that they should not lack hope when seeing all the social and environmental challenges around the world.
- Spend time in class explaining what is a good or a not-so-good essay, provide examples of different actions the students could take and explain to them that any small action counts towards supporting the SDGs. Indeed, a few students seemed to be lost when thinking about how they could complete this assignment and several of them had the feeling at the beginning that they could not do anything to support the SDGs.
- The linkages of one SDG to the others are quite obvious (see question 5 of the assignment in Table 11.1). They can all be linked to one another, and several students identified that. This last question could be deleted or formulated in a way that further encourages students to share their experiences and reflect on their learning.

Overall, I would encourage any teacher, but also researchers and consultants, to experiment further with the power of emotional learning. I did not know of this concept before developing the exercise; I just thought: "given the urge to do something against climate change and other environmental and social issues, what can I do in my lecture that will have an impact?" It is possible to give a lecture with the highest motivation and energy, but this will never have the same impact as turning students into change actors. This experience not only had an impact on the students, but also on me. Several essays made me cry and then I thought: "I managed to reach students' heart, to make them think critically, to open their eyes to a systems perspective, a holistic understanding of sustainability issues, I managed to make them realize that they can change things and that it starts with an introspection about one's own sustainability mindset." I have never felt so satisfied and proud of the content of one of my courses and its impact, and I am really eager to replicate this exercise and develop new ones from October 2020! I am very grateful to all my colleagues from whom I learned about the sustainability mindset. Now I feel I have found a very effective way to "shape the world [I] want to live in" that is highly complementary to my research work. With this chapter, I would like to encourage all researchers, especially in the management sciences where contacts with our main target (firms) is not always easy, to engage in lecturing and training students to contribute to their development as managers with a sustainability mindset. Indeed, our targets should not only be practitioners, but also students. After this experience, one of my favorite quotations, which is from Nelson Mandela, appears to be so true and genuine: "Education is the most powerful weapon which you can use to change the world."

## Notes

1 https://www.excelia-group.com/live-unique-experiences/humacitec.
2 https://refedd.org/sedd/.

## References

Asah, S.T., & Singh, N.C. (2020). SEL for SDGs: Why social and emotional learning is necessary to achieve the Sustainable Development Goals. https://mgiep.unesco.org/article/sel-for-sdgs-why-social-and-emotional-learning-is-necessary-to-achieve-the-sustainable-development-goals

Festinger, L. (1957). *A Theory of Cognitive Dissonance* (Vol. 2). Stanford University Press.

Ives, C.D., Freeth, R., & Fischer, J. (2020). Inside-out sustainability: The neglect of inner worlds. *Ambio*, *49*(1), 208–217. https://doi.org/10.1007/s13280-019-01187-w

Rimanoczy, I. (2016). *Stop Teaching: Principles and Practices for Responsible Management Education*. New York: Business Expert Press.

Rimanoczy, I. (2017). *Big Bang Being: Developing the Sustainability Mindset* (2nd ed.). New York: Routledge.

Rimanoczy, I. (2020). *The Sustainability Mindset Principles: A Guide to Developing a Mindset for a Better World*. London: Routledge.

Shrivastava, P. (2010). Pedagogy of passion for sustainability. *Academy of Management Learning & Education*, *9*(3), 443–455.

United Nations. (2015). *Transforming Our World: The 2030 Agenda for Sustainable Development*. https://sustainabledevelopment.un.org/post2015/transformingourworld/publication

# 12

# CLIMATE CHANGE IN ACTION

## Creating innovation and increasing impact in undergraduate business curriculum design

*Ayako Huang*

## Introduction

The notion of integrating sustainability in management education has witnessed a dramatic increase over the last decade, resulting in a broad range of studies published about pedagogical strategies. The problem of scholarship in this area is the lack of accountability and efficiency in measuring learning outcomes. The significance of pedagogy reform is now widely recognized.

A further important issue that may well be elucidated by academia is the definition of sustainability itself. Sustainability refers to a deep cultural understanding of living within natural systems as well as a deep spiritual connection to the earth and others (Capra, 2002; Macy & Brown, 1998). Sustainability education has become evident in educational interventions to build skills in developing empathy as well as promoting global citizenship. The transformation in sustainability management education therefore requires addressing mindsets as part of a deep inner development.

This idea can be successfully integrated with the Head, Heart and Hands model introduced by Orr (1992) and expanded by Sipos, Battisti and Grimm (2008). The model shows the holistic nature of a transformative experience and links the cognitive (Head), the affective (Heart) and the behavioral (Hands) domains (Singleton, 2015).

In this new reality, pedagogical approaches that consider the Head, Heart and Hands model are often used in sustainability management education. Students are trained in Head and Hands approaches to immediately measure the impact of the knowledge and skills that they will need in a business context – for example, to measure the degree of transparency in stakeholder engagement. Yet, the spiritual and emotion-related elements from the Heart approach to teaching are nearly absent. More specifically, developing students' empathy, mindfulness and using social learning to reflect on their impact in a social context is narrowly confined to sustainability courses in management education.

This chapter takes a descriptive stance and an in-depth look at the implementation of the sustainability pedagogy of one undergraduate business course. The course highlights the importance of sustainability pedagogy to raise awareness and promote behavioral change.

DOI: 10.4324/9781003229735-14

The following provides a theoretical outline of the learning model used in this chapter as a new, instructive way to enrich the whole self and encourage personal connection around sustainability, as well as a discussion of implications for the implementation of self-learning sustainability pedagogy in management education settings.

## The Three Phases model

The Head, Heart and Hands model in sustainability education supports changes that go beyond behavioral change, as well as in challenging existing beliefs, feelings and ideas. Stemming from this thought, sustainability pedagogies are increasingly in favor of offering a value-added approach as key for sustainability transformation.

Inviting students to explore their feelings and experience their connection with nature are noteworthy strategies to show students how to develop value and social systems that are resilient and sustainable. This requires one to scrutinize one's own beliefs, attitude, interactions, to accelerate the learning process. Kassel, Rimanoczy and Mitchell (2016) further define such a process as "Sustainability Mindset," which entails personal development to create a positive, significant, social impact in the world.

Social impact refers to the degree of changes to individual attitudes, values, beliefs and behavior that contribute to wellbeing (O'Flaherty & Liddy, 2017). More specifically, a fundamental shift in how students think and act enables them to make behavioral change. Yet, the vast majority of sustainability scholars are still focused on outcome-based learning and believe changing behaviors results in a values shift. For some, sustainability management education is viewed as measured on performance rather than process development.

In line with the above notion, the Head, Heart and Hands model does not include the process of development in the connectedness, for example, how an individual thinks, feels and engages with nature. As a result, the focus of most sustainability pedagogies over recent decades has remained a hands-on experience of the solutions to the climate crisis. Noddings (1997) argues that a "morally defensible aim for education ... should be to encourage the growth of competent, caring, loving, and lovable people" (p.184). If we want to reconcile the HHH model with a shift in beliefs and values, we need a different pedagogical approach.

This chapter proposes a values-based learning approach for the management discipline, and especially addresses the assessment and evaluation of both impact and sustainability mindset factors based on the Three Phases model. Table 12.1 summarizes a theoretical construct and measurement of the social impact in the "Three Phase model."

Conceptually, this model begins with the first phase "Head," which puts an emphasis on comprehension of knowledge and skill to increase students' systemic understanding of complex sustainability issues. With this systemic understanding, students can make inferences about beliefs and attitudes related to the responsibility of oneself to the ecosystem (Orr, 2004), as shown in the impact indicators. This phase, thus, is ingrained in systems theory (Capra, 2002; Meadows, 2008).

The second phase, "Heart," underlines the connectedness of one's thinking, feelings, intuition and sensation of nature, as well as social sensitivity, resulting in the expansion of experiential value (Pugh, 2002) and the enhancement of motivation and engagement (Strong, Silver & Robinson, 1995). Outdoor activities in this phase rely primarily on experiential learning theory (Horlings, Nieto-Romero, Pisters & Soini, 2020; Kolb, 1984). The outdoor education

**TABLE 12.1** The Three Phases model

| *Three Phases* | *Learning theories* | *Learning activities* | *The impact indicators* |
|---|---|---|---|
| The first phase – comprehension (Head) | system learning | -dialogue and discussion<br>-display board<br>-field trip | understand sustainability and create personal behavior changes |
| The second phase – connectedness (Heart) | experiential learning | -outdoor classroom<br>-infographic project | reflect emotional connection and commitment to sustainable projects |
| The third phase – competence (Hands) | project-based learning | -consultant project | build competence, application skills, and shift relational values in long-term practices |

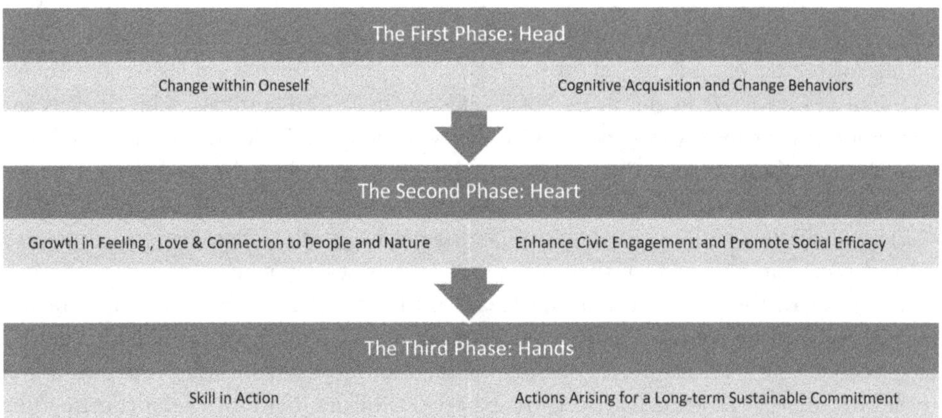


**FIGURE 12.1** The sustainability mindset development indicators diagram

allows for a personal reflection about sustainability and promotes ecological consciousness, which can develop new commitments, constituting the second phase impact indicators.

The third phase, "Hands," incorporates the competence to act as the response to the "Heart" phase. This phase is based on capacity building, awareness, participation and action strategies. It consistently emphasizes activity derived from project-based learning theory (Dewey, 1938). Students are assigned a hands-on project, where they design and implement sustainability activities and communications about climate change. The central notion guiding the impact indicators in this phase is the competence of applying what they learned and skills to lead long-term sustainability actions.

As noted in the above section, a starting point for developing the sustainability mindset is framed by the cognitive, socio-emotional and behavioral process. Figure 12.1 illustrates a value shift in the development of a sustainability mindset involving both *time* (the first phase, change within oneself; the second phase, the effects of a commitment to influence others; the third phase, skill in action to implement changes), as well as *place* (multiple levels of exposure in nature and local communities). Mindset development depends on the personal thoughts and experience, which color observation and perception, modifying beliefs and attitudes. And naturally, this reflects in one's actions. To this extent, mindset development is the central component of a sequential and subjective learning process. Thus, disciplinary boundaries should be set aside; learning sustainability in management education is an active and reflective construction process of change, refinement and transformation within the learner-centered domain.

## Context

This section provides a description of an undergraduate Management and Organization course based on the Three Phases model, to highlight the importance of a sustainability learning approach that is systemic, interdisciplinary and experiential.

The undergraduate students were enrolled in an introductory Management and Organization course at a small Midwestern university. The course, offered every fall semester, meets for four weeks Monday through Friday 10:00 am to 3:10 pm, and Saturday 10:00 am to 12:00 pm, including a daily group meditation of 20 minutes, for a total of four credits. Students represented business majors as well as non-business majors due to the course's fulfillment of several general education requirements for the university.

This course had no stated connection to prior sustainability learning and thus provided an example of how sustainability pedagogy can be implemented within the management discipline. It incorporated both management and sustainability into a curriculum around the theme "climate change action and management." It was specifically redesigned to incorporate learning theory and the Three Phases model.

The course included several major assignments: a carbon footprint analysis report, an infographic project and a circular economy consultant project. All these assignments were related to the course theme. Another assignment included a reflection journal, which required students to write a page in their reflection journal based on both readings and content of each lesson.

Daily dialogue and group discussion sessions were led by students and consultant clients. Field trips were also incorporated into this course in order to provide contextual and experiential learning about local sustainability.

## The first phase – comprehension (Head)

Sipos et al. (2008) define "Head" as the cognitive domain – to learn, to understand and to comprehend the ecological and sustainability concepts. The first phase includes an expansion of perception and experiential value through different pedagogies in order to engage and motivate students. The "Head" phase relates to climate change education by teaching the facts and providing information about climate change to engage students in the issue and help them understand the reason behind climate change projects or initiatives. Facts can be a catalyst for students to consider their impact on the natural environment.

This class began with each student showing one climate change impact photo of their choice. A class dialogue followed in which students were required to share their feelings and concerns about natural systems and to meld cognitive skills with empathy for all life. The course instructor broadly defined the topic of the dialogue first, but students were encouraged to come up with their own questions for the dialogues.

On the same day, students were asked to complete a carbon emission analysis report as the homework assignment. Students wrote down the implications of their personal carbon footprint. What would they change to lessen their impact? Students also listed the pros and cons of each of their changes. What is the likelihood that they actually will make these changes to be more sustainable? The report was intended to create a rational analysis to facilitate their efforts and prompt them to live more sustainably.

Students were encouraged to participate in a local climate change rally event and schedule an interview with climate change activists during the first week of the class. The purpose of the interview was for students to understand environmental challenges and the negative impacts that these challenges exerted on the environment.

The following weeks students prepared a 15-minute brainstorming session to discuss and determine six climate change headings using big blank posters around the room. The title of headings included: Future Impact of Climate Change; Role of Individuals; Role of Business Leaders; Role of Government; Role of Local Community; and Role of University. Once students confirmed their final headings, they spent 20 minutes writing down their thoughts under each heading.

After they completed writing their thoughts, one student began facilitating a group discussion to address each heading. Since they had done the carbon emission analysis report, participated in the local climate change event and read the assigned articles, the discussion flowed more naturally into the subject, "How to promote social awareness and seek institutional changes that embed a promising future for nature in the local community?"

The pedagogy in the "Head" phase focused on the comprehension of sustainability concepts with personal choices by encouraging students to "shape their world" through their physical and social connections with the local environment. Writing a carbon emission analysis report required that students look deeply at their own contribution to GHG emissions. Interviewing the environmental activists gave an opportunity for students to examine their own biases *vis-à-vis* a worldview on climate change issues. A group discussion helped students learn to evaluate their thinking skills and to communicate their thoughts. More importantly, students recognized their role in climate change and recognized that they are important and "part of the solution" for climate change. Ultimately, it created a stronger sense of connectedness with their community as they passionately practiced their fresh thinking about grassroots initiatives.

The "Head" phase is most closely related to systems theory. The theory provides a broader perspective to understand human–nature interactions connected with complex sustainability issues (Holling, 2001). It also calls for an interdisciplinary pedagogy, including academic and scientific expertise from stakeholders to discuss both scientific and societal problems (Jahn, Bergmann & Keil, 2012). As such, the theory stimulates students to study climate change issues from a resilience-thinking perspective.

### *Impact assessment of the "Head" phase: Change within oneself*

Chapin, Sala and Huber-Sannwald (2001) suggest that through a series of feedback system students can expand the notion of "reciprocal influences" among sustainability contexts.

We measure the impact of sustainability education through skills such as emotional, intellectual and social awareness and change. In this phase, students are assessed on the degree of their new understanding of sustainability through their reflection and self-awareness, as well as their recognition of the gap between espoused values and personal values – both required to change behaviors (Rimanoczy, 2013).

Students expressed that through this phase they came to understand the concept of sustainability more holistically. The pedagogies provided a multi-dimensional understanding of sustainability through the climate change advocates they interviewed, the local event in which they participated, the group discussions they shared and the reading assignments – all of which were related to the course theme. Specifically, students became more conscious of their awareness with respect to the impact of climate change, and adapted their behaviors accordingly. They recognized that any sustainability initiatives begin with managing oneself.

In their carbon footprint analysis report, almost all students indicated increasing self-awareness of climate change impact and, more concretely, were willing to change their behaviors to reduce GHG emissions.[1]

One student wrote:

> *The carbon footprint analysis report shows me that we need to work to get more people to commit to CO2 neutrality. I plan to implement this change immediately.*

Following the same thought, another student responded to the previous statement:

> *I will change my little habits to prove the importance of making small changes to make a difference in sustainability.*

In this regard, most of the students stated how to change their personal behaviors in supporting sustainability pathways. For example, one student acknowledged how to create the change:

> *There's no doubt climate change is changing the way we live ... My current emission of waste is 24,493 lbs. I would like to start composting in my garden. I feel like if I do that the excess food won't go to waste, I could nurture my garden and it will make me feel better about my impact on the planet.*

Comprehending the concept of sustainability with embodied experiences leads to a new social consciousness to change on a personal level. Another student mentioned about this change:

> *I have a clearer understanding of how the themes of sustainability are actually interconnected with my life ... I will try not to toss the clothes in the dryer on sunny days.*

More broadly, students came to understand sustainability more holistically and concretely, primarily through their learning experience. One student formulated the notion of "collective values" in his report:

> *Things we can do better to reduce our carbon footprint:*
>
> 1. *Encourage our friends and family to take the Carbon Footprint Analysis and make changes in their habits to reduce their carbon footprints.*

2. *If we can increase our impact by [10] people, and they each increase their impact by 10 people, and so forth and so on – now that's making an exponential difference.*

Another student shared a similar thought, stating:

> *When establishing our own mind by educating ourselves and being proactive in our voice, we can be green representatives for the people around us who look up to our decisions and will enact their own values just by learning from others. This ripple effect is a real consideration when deepening our awareness on how we can impact our global climate.*

Students reflected on their roles and positions toward sustainability and social awareness while participating in class activities. They defined the meaning of sustainability from an expression of their life experience and values, acknowledging the reciprocal relationship between behavioral change and a sustainable future. In parallel with the systems learning approach, students understood sustainability is inherent in environmental, social and economic phenomena, which are all interconnected. Attention to the learning process is important as it affects how students shift their mindset and learn to take responsibility to act on climate change.

## The second phase – connectedness (Heart)

The second phase, connectedness to Heart, is the phase where emotion lives. The "Heart" phase refers to the affective domain where students enable their new values and attitudes and translate them into behaviors. Wangaard, Elias and Fink (2014) note that meaningful learning requires affective engagement (Heart) to create a learning experience in order to achieve cognitive mastery (Head). Facilitating students' efforts to connect their feelings to nature and to others therefore requires a process and is participatory, experiential and relational. This phase relies primarily on experiential learning theory (Dewey, 1938; Kolb, 1984).

Experiential learning refers to the learning process based on the principle of "learning by doing" (Kotti, 2008). Acquisition of knowledge is strongly linked with new experience or doing things to which one is personally connected. A connection with nature is often a subjective and personal experience. Facilitating students to connect to nature is a critical learning experience to motivate them to foster their values and the ecological paradigm. Value is part of the affective system and relates to affective thoughts that guide behaviors (Rokeach, 1973). In other words, feeling is the fundamental reason for actions. Experiencing the beauty of nature is critical to develop values and systems thinking to understand the complexity and interconnections toward sustainability.

The primary aim in the "Heart" phase is to raise a connectedness with nature and others. The pedagogy started with developing a connection with nature and the role of a change agent in climate change initiatives. Exploring with fun and innovative ideas to find solutions to reduce the local carbon footprint was the second pedagogy approach in this phase. The phase ended with a facilitation process for the commitment to a campus consultant project.

### *Connectedness*

The second week of the class, the pedagogy focused on helping students experience an emotional connection with nature and appreciate its value. One of the class activities is to learn about campus ecology, "Exploring and Knowing Your Campus." In doing so, some class sections were taught in the amphitheater when weather permitted. In addition, students were asked to explore the campus and find their favorite tree. They sat an hour under that tree in silence reading the assigned articles in three afternoon sections. By then, students had written and illustrated journal entries about the change of their relationship with nature, after spending some time quietly observing and experiencing a natural setting of their choice.

Through experience with nature, students found personal relevance in exploring the campus ecology, adding purpose and value to their learning. This educational experience brought a new perspective of relationship and responsibility between students and the local biosphere (Tooth, 2018). When students understand that everyone is an interconnected part of nature and learn how ecosystems support their lives, they are able to develop caring attitudes and expand their values, which motivate behavior change.

## *Exploration*

When discussing the "Heart" phase in relation to climate change, some scholars have suggested that learning needs to be exploratory, empowering, fun and creative (Barthel et al., 2018; Chawla, 2007), shining a light on the inner dimension of emotions. Embracing unexpected discussions may lead to developing an exploratory process. Once the "Heart" piece is in place and students have refined their thoughts, an open discussion becomes more palatable and dynamic. Such an approach can foster critical reflection and cultivate new practices to engage with the new climate change activities.

After experiencing nature on campus, students began to appreciate and express their feelings about the campus by questioning themselves: "What can we do to reduce the impact of climate change on campus?" Through a group discussion, students were able to move closer to understanding the broader social purpose of their role, not only in contributing to reducing the impact of climate change, but also to help others by making changes.

During the discussion, students were eager to conduct a climate change consultant project, which led them to reflect upon their impact on campus and the value of reducing their ecological footprint. Moreover, the project focused on describing and possibly altering the behaviors of campus staff and other students in order to support the carbon neutrality goal of the climate change project. Students continued to explore what it means to change behavior by engaging in group dialogues with follow-up reflection questions. For example, how do human activities in the café and deli contribute to the carbon footprint? What is necessary and what is possible in order to reduce the carbon footprint of both shops?

As a result, students volunteered for an infographic project, dividing into two groups – one for the café and one for the deli shop. The purpose of the project was to educate and promote the importance of carbon footprint with mitigation strategies. The infographic-art project allowed students to develop information literacy skills and analyze scientific data, and involved them in graphic design. This project empowered students to make informed decisions that would positively impact on-campus life. The infographics were later posted physically on campus as well as electronically on the university's app. The content of the infographic included the implications of certain types of environmental impacts relating to campus GHG emissions. Examples of "environmental emissions" included carbon footprint, waste management, etc. The grading criteria for the infographic were as follows:

1. Descriptive of the science in language appropriate for campus students, staff and faculty
2. Informative as to the wider implications
3. Offering productive ideas for individual action

### *Commitment*

Sipos et al. (2008) highlight that the affective domain (Heart) is the foundation of a value-focused approach, encouraging students to develop a meaningful collaboration to create better and more appealing alternatives and sustainable solutions. This school of thought defines motivation, attitude and commitment as key elements for a transformational learning experience.

The infographic project was a meaningful way for students to connect with purpose, in their own lives and beyond. It provided them a participatory approach to explore their sense of place with the aim of building a place narrative to reduce carbon footprint. One such attempt was to foster social awareness among students. As part of the infographic project, students walked on campus and spoke to their friends to better understand their perspectives and attitudes. They participated in voluntary meetings to discuss how to further promote this project. The form and content of the promotion were left to the discretion of students, but all scheduled group advisements with the course instructor to discuss each promotion plan in its early stages. Becoming more conscious of carbon footprint with respect to campus life, students adapted their research accordingly.

The infographic project was a collaborative, inquiry-based and self-directed learning initiative. Students were highly committed to a learning opportunity in which they were innovative designers as well as global collaborators.

### *Impact assessment of the "Heart" phase: Connect to nature*

The "Heart" phase is based on students' experience with nature in order to construct an emotional connection as well as to cultivate the ambition to contribute to sustainability change initiatives. The impact assessment for this phase is built on how students reflected their viewpoints in their social responsibility actions and promoted change in the place they live.

The effects of psychological and socio-ecological processes can serve as a connector for climate change learning (Schweizer, Davis & Thompson, 2013). Related to the above, one student described the process of the personal transformation to a commitment to climate change action:

> *Now, I understand myself better. I can connect to nature in a personal way. This deepens my understanding of nature … for every action there is a reaction. Every time I reuse or repurpose, I reduce waste. Reducing waste/managing resources is good for the environment and economy.*

Another student shared her comment that "*an infographic serves as a voice in the recognition [that] an individual can be a part of a collective society where every action matters.*"

Besides the personal contribution mentioned above, emotions link to one's inner values and inspire the actions necessary to engage in promoting climate protection. One of the students from the café group, who started off with the ambition of being a change agent and starting a discussion, raised awareness with Student Government representatives, the Student Life department, and the café manager to promote a "bring your own mug" campaign. As experienced by this student: "*When I think of my project and work, it allows me to share the important messages … makes me feel I can do something meaningful for others.*"

Being able to anticipate and assess the impact of personal and local community involvement through sustainable projects, students explored social, cultural and political systems in order to transmit their beliefs and attitudes. Related to this, one student noted: *"How we think, what we do, how we listen, what we say, etc., ALL have a part in us being in our economy."* One student became even more conscious of the public policy issue and critiqued:

> *My takeaway from today's lesson is we need to work to get the U.S. and other world powers to commit to this kind of dedication to CO2 neutrality. If we don't get CO2 neutral, not much of the rest matters. We need to begin it with our own place.*

The infographic project for the "Heart" phase also connected the sustainability and management domains of knowledge and developed in the students a certain emotional relationship with their locality that facilitated certain change actions. In this learning cycle, students apprehended and experienced both sustainability and management principles through a reflective observation. In other words, this learning cycle provided insights into the emotional factors of what students needed to do with their managerial skills in order to transform unsustainable societies into sustainable societies. Therefore, sustainability education was based upon the vision of students.

When students began to observe what a sustainable campus means, some of them explored the variety of positions they would take as a leader and what role would suit them best to work on promoting the infographic project. One student expressed: "*I want to do things differently … to influence my community through the collaboration with others. This is the accountability of being a true leader.*"

Another student analyzed the infographic project in the broader context that any climate change actions need to be based on consensus of both the political and academic stakeholders. The student shared her experience:

> *The infographic project is a deal of organizational culture and bureaucratic power which makes this project possible … . We need to change the mind of key players to love and appreciate the nature of this place.*

Aligning with experiential learning theory, students learned about socio-ecological impact through their concrete infographic experience and conceptualized the term "sustainability" with their own reflective observations. Every feeling, thought and action is constructed in a deep "Heart" level. Students started with a commitment to support change towards sustainability on campus. This commitment is the internal ("Heart") beginning of what is manifested in the external students' sustainability activities: thus, the action emerged from the "Heart."

The fact that students considered themselves as being part of the campus translated intention into social responsibility, thereby contributing to change in the place they live. Experiencing the awareness of connectedness to the place can change the relation between people and their environment (Horlings, 2019). It is evident that understanding nature is a new orientation for integrating social and ecological concerns into the teaching context

## The third phase – competence (Hands)

The "Hands" phase refers to the ability to perform skills with accuracy and competency, including precision, coordination and manipulation (Jensen, Christina & Geraldi, 2016). This phase is especially relevant in the context of a project-based learning approach. When students invest time in an authentic project, they are required to explore topic with different stakeholders, apply knowledge and use analytical skills to learn from real-world experiences. Such a learning process drives changes in how they collaborate, how they relate to others and in other and daily habits.

In this phase, the assessment indicators are mainly based on sustainability competencies – students need to "link [a] complex set of knowledge, skills, and attitudes that enable successful task performance and problem solving with respect to real-world sustainability problems, challenges, and opportunities" (Wiek, Withycombe & Redman, 2011: 204). Additionally, the assessment indicators include students' abilities to share their concerns about the future and to find their place in a long-term commitment for climate change.

In the beginning of the third week, students proposed to follow up the infographic project by conducting a consulting project to convert both the café and the deli shop into a circular economy model. Over the following two weeks, students were encouraged to schedule a site visit and to interview shop managers. During this time, students also researched sustainable strategies that could be beneficial in addressing the topics in which the managers had indicated an interest. At the end of the third week, students compiled their research findings and in a mid-term reflection presentation offered three options for moving forward.

The main aim of this presentation was to assess their application of class concepts and reflect upon the team's performance, in order to prepare any clarifications or fuller explanations that might be needed for the final presentation. The rest of the course was spent working on interviewing stakeholders and analyzing a financial report for the project. During this time, the stakeholders were especially helpful when it came to the financial analysis and campus policies. The shop managers suggested the students calculate return on investments and be cognizant of campus regulations and policies. At the close of the class, students gave a 30-minute professional presentation to the shop managers with the recommended mitigation strategies and the respective cost-benefit analyses.

To monitor changes in the outcome of both projects, the course instructor tracked changes over time after the completion of the class to evaluate whether the project had a continuous impact on the campus. The café shop implemented the "bring your own mug" campaign and agreed to source only locally as a result of the consulting.

Furthermore, the manager recognized the need for fostering sustainable behaviors. Critical to this agenda was strengthening global solidarity through continuing to promote mitigation strategies.

## *Impact assessment of the "Hands" phase – skills in action*

The "Hands" phase allowed students to go beyond traditional lectures and design a hands-on experience for their commitment to make an impact on campus. The impact assessment in this phase first reviews the application skill that links environmental sustainability and management principles. The second part of the assessment focuses on an increasing sense of responsibility to commit to a long-term sustainable future on personal, social and societal levels.

In this context, the assessment indicators evaluate the application of management concepts as the first criteria for implementing sustainable projects. This type of assessment aims to recognize how the class project aligned with management concepts that emerged from one's inner development. As noted by one student: "*The nature of life is to grow. I can see from this reflection journal how much I have grown over the last five years, while here at MIU.*"

Becoming more conscious in relation to personal awareness and the learning framework, students were able to apply their self-management learning in a more profound way. It changed the way the students defined leadership. The following illustrates an example from a student:

> *The most important thing I learned from this project is if I want to influence my [group] members to lead in their own lives, I must also lead in my life. Being a good leader takes knowing yourself to know the people you want to lead.*

Any sustainable project comprises multi-sectorial and multi-stakeholder partnerships to promote synergies and reach common goals. Teamwork therefore ensures that a sustainability project can be done successfully. One student expressed the benefit of working with the team:

> *I like to work in teams that create friendly and lasting relationships. For example, [in] the last presentation of our deli project, I learned the process of implementing a circular economy from a much broader perspective.*

The second assessment of the impact of the "Hands" phase focuses on the process of raising ecological consciousness through reflection to promote environmental stewardship.

Related to this, students noted specific examples of their transformation experience in describing their understanding of sustainability, acknowledging the importance of creating an impact on campus. One student asserted:

> *I felt excited and happy about the project. I felt like we could really have a positive impact on the university and help the café save money, while increasing sustainable awareness on campus. It felt like I was finally putting my BA in Sustainable Living to work, and that felt good.*

Taken together, the class project helped students face their emotions in a constructive way and that facilitated their responsibility for future change. As one student said in the following comment:

> *This learning experience was significant to me as now I have gained more knowledge and skill; I can be more productive in group projects and make decisions in climate change initiatives.*

Tapping into a quotation from the founder of Maharishi International University – "Knowledge is for action. Action is for achievement. Achievement is for fulfillment" (Maharishi Mahesh Yogi, 1974) – when students established a sustainable vision in the consultant project, they enjoyed the sense of fulfillment from having achieved their desired outcome. One student shared her feelings:

> *I find it very important to work on our café project. We want to promote using mugs. The cup discount is important to me because I feel very strongly about reusable cups and how they affect the environment. I am happy to know this discount promotion will be adapted in the café shop.*

In addition, students understood that climate change was not simply a "class event." It was a holistic and long-term change in thinking and actions in order to build a sustainable future. One student desired to continue playing a role in preparing societies for that movement:

> *After our final presentation, my values lie strongly in sustainability and I want to work for companies that are building a better world … . I want to work for companies that already have roots in sustainability in hopes to reverse some of the effects of climate change.*

A project-based learning approach plays a paramount role in helping students to develop mitigation skills, at the same time raising awareness about climate crisis solutions. Teaching sustainability in management education should focus not only on the application skills to challenge the reality of the issue of climate change, but also on enriching the mindset development to motivate the change effort.

Mindset development is an internal process that requires time. Alongside this process, the social impact from sustainable projects may also involve long-term engagement rather than immediate measurable results (O'Flaherty & Liddy, 2017). Monitoring and evaluating impact continue to be a challenge. New evaluative schemes should be created, including competence building and a self-awareness component. To achieve this, the curriculum needs to help students connect to life experience, explore the meaning of life, and build the competences required for action to address sustainability risks.

## Implication

There is a disconnect between environmentally related beliefs, emotions and meanings (Maiteny, 2002) as well as attitudes and behaviors (Kollmuss &Agyeman, 2002; Marcinkowski & Reid, 2019) in the current education system. The methods to address such conundrums have to reform the domains of educational purpose from an overemphasis on the acquisition of knowledge (Head) and application of skill (Hands) to more of a focus on the transformation of norms and values by the connection of oneself with nature and others (Heart).

This learning pedagogy begins with developing students' values and ecological consciousness, including understanding the reciprocal influences of a variety of sustainability contexts. The purpose of the first phase is to make sure students can not only generate and acquire new knowledge, but also reflect on their own behavior and values. The second phase offers students the opportunity to create a personal bond with the environment as a "relationality" approach, which fosters students' responsibility through promoting collective awareness in global sustainability. The last phase focuses on the tasks at hand by providing a hands-on experience project for students to interconnect their own individual sustainability contexts and apply their skills to implement community and global sustainability.

The Three Phases model tries to measure the transformation of the way students think and act. It involves the development of a personal mission and social sensitivity (Rimanoczy, 2013; Hermes & Rimanoczy, 2018) through a sequence of class projects aligned with learning objectives within the phases. As such, the aim of this model is to promote learning processes within certain contexts rather than focusing on learning outcomes with an analytical approach.

A value-based approach is a particular type of pedagogy in which students take a critical and explicitly normative perspective in relation to nature, and engage in social awareness as an outcome of sustainable mindset development. It includes a shift of one's mindset from micro to macro scales, from short to longer-term commitments, and from single to multidisciplinary studies (Rimanoczy, 2013; Hermes & Rimanoczy, 2018). However, a number of challenges have arisen from this chapter notably, one questions how to best define the purpose of education and identify learning outcomes. As pointed out by Freire (1972, p. 53), "liberating education consists in acts of cognition, not transferals of information." The potential for the change of mindset is limited in the existing educational system. The main confrontation is that learning outcomes are often predetermined and also are limited to the process of learning and development of self (O'Flaherty & Liddy, 2017).

To introduce a sustainable mindset development into educational learning sites is challenging yet necessary; university curricula need to offer comprehensive and systematic class activities through the Head, Heart and Hands model to develop an understanding of sustainability, as well as to learn to live sustainably and to take responsibility. In this way, any pedagogies cannot be separated by the "cognitive," "affective" and "behavioral" phases. The learning approach needs to include the values component to engage in deliberative discussions, reflective reasonings and practical projects as the basis for developing a sustainable mindset through the three phases.

Therein, the key to the Three Phases model is namely:

- A systematic development of students' climate change awareness and knowledge
- An interactive, learner-centered pedagogy with students' significant life experience, motivation and values
- An innovative measurement of sustainability to ensure a mindset development rather than actions
- A student-empowered approach for students to take ownership over the projects, set goals, monitor progress, make adjustments, reflect on their learning and celebrate their progress and accomplishments.

Sustainability cannot be simplistically taught by the fragmentation of learning domains. Teaching students to comprehend environmental knowledge does not lead them to changes in their attitudes or behaviors. Moreover, reflecting an understanding of eco-literacy is not necessarily a guarantee that students will engage in any hands-on projects. A constructive pedagogy must include a refinement to sharpen students' minds, to illuminate the connections with broader spheres and to lead students in their commitment to a sustainable lifestyle.

## Conclusion

Sustainability management education is a human-centered and multi-dimensional task with a systematic approach. Establishing recommendations for a cross-curriculum model through Head, Heart and Hands learning contexts could result in more holistic educational outcomes.

As noted earlier, the role of sustainability education is in "changing minds, not the climate" (UNESCO, 2017). This vision of wishes, commitment and responsibilities in relation to pedagogy is embedded in the Heart phase such as sense-making and self-transformation. Surely, there is a gap between the Head and Hands phases. The vast majority of scholars are either overstretched in the Head phase – theoretical concepts – or overloaded in "Hands" – practical skills. More importantly, there is no holistic approach to monitor the impact of the sustainable mindset development process. The formal education setting has noted that the learning outcomes cannot be predefined, which raises difficulties for assessing this impact (O'Flaherty & Liddy, 2017).

Often, scholars lead class activities or class projects related to one specific phase without integrating them with any other phase's learning objectives and impact assessments. Such disconnects call for an urgent change for Head, Heart and Hands pedagogy design in sustainability education.

To grasp the original statement from Brubacher and Rudy (1976, p. 268), "true education was the education of the self." This education of the self is dependent upon a shift to understand the nature and relationship with oneself and a new system of thought that integrates social and ecological concerns. This is a systems approach to "mindset" development. Quite clearly, the reform of sustainability management education is necessary to provide a learning context with holistic pedagogical strategies to highlight the mindset shift process within and beyond the classroom setting.

## Note

1 All student voices were used with permission and were anonymous for this paper

## References

Barthel, S., Belton, S., Raymond, C.M., & Giusti, M. (2018). Fostering children's connection to nature through authentic situations: The case of saving salamanders at school. *Frontiers in Psychology, 8*. http://doi.org/10.3389/fpsyg.2018.00928.

Brubacher, J., & Rudy, W. (1976). *Higher Education in Transition: A History of American Colleges and Universities*. New York: Harper and Row.

Capra, F. (2002). *The Hidden Connections: A Science for Sustainable Living*. New York: Anchor.

Chapin, F., Sala, O., & Huber-Sannwald, E. (2001). *Global Bio-diversity in a Changing Environment*. New York: Springer-Verlag.

Chawla, L. (2007). Childhood experiences associated with care for the nature world: A theoretical framework for empirical result. *Children, Youth and Environments, 17*(4), 144–170.

Dewey, J. (1938). *Experience and Education*. New York: Collier Books.

Freire, P. (1972). *Pedagogy of the Oppressed*. Harmondsworth: Penguin.

Hermes, J., & Rimanoczy, I. (2018). Deep learning for a sustainability mindset. *The International Journal of Management Education, 16*(3), 460–467.

Holling, C.S. (2001). Understanding the complexity of economic, ecological, and social systems. *Ecosystems, 4*(5), 390–405.

Horlings L.G. (2019) Sustainable Place-shaping: What, Why and How. Findings of the SUSPLACE Programme; Deliverable D7.6 Synthesis Report. Rural Sociology Group, Wageningen University and Research, Wageningen. Retrieved from https://doi.org/10.18174 /50166 6.

Horlings, L.G., Nieto-Romero, M., Pisters, S., & Soini, K. (2020). Operationalising transformative sustainability science through place-based research: The role of researchers. *Sustainability Science*, 15, 467–484.

Jahn, T., Bergmann, M., & Keil, F. (2012). Transdisciplinary: Between mainstreaming and marginalization. *Ecological Economics, 79*, 1–10.

Jenkins, A, Christina, T., & Geraldi, J. (2016). The projectification of everything. Projects as a human condition. *Project Management Journal, 47*(3), 21–34.

Kassel, K., Rimanoczy, I., & Mitchell, S. (2016, August). *The Sustainable Mindset: Connecting Being, Thinking, and Doing in Management Education*. Paper presented at the 76th Academy of Management, Anaheim, CA.

Kolb, D.A. (1984). *Experiential Learning: Experience as the Source of Learning and Development.* Englewood Cliffs, NJ: Prentice-Hall.

Kollmuss, A., & Agyeman, J. (2002). Mind the gap: Why do people act environmentally and what are the barriers to pro-environmental behavior? *Environmental Education Research, 8*(3), 239–260.

Kotti, D. (2008). Experiential learning from theory to practice. *Adult Education, 13*, 35–41.

Marcinkowski, T., & Reid, A. (2019). Reviews of research on the attitude–behavior relationship and their implications for future environmental education research. *Environmental Education Research, 25*, 459–71. https://www.tandfonline.com/doi/full/10.1080/13504622.2019.1634237

Macy, J., & Brown, M.Y. (1998). *Coming Back to Life: Practices to Reconnect Our Lives, Our World.* Gabriola, BC: New Society.

Maharishi International University. (1974). *Maharishi International University Catalogue, 1974/5.* Fairfield, IA: MIU Press.

Maiteny, P.T. (2002). Mind in the gap: Summary of research exploring "inner" influences on pro-sustainability learning and behavior. *Environmental Education Research, 8*(3), 299–306.

Meadows, D. (2008). *Thinking in Systems.* Vermont: Chelsea Green.

Noddings, N. (1997). Thinking about standards. *Phi Delta Kappa, 79*(3), 184–189.

O'Flaherty, J. & Liddy, M. (2017). The impact of development education and education for sustainable development interventions: A synthesis of the research. *Environmental Education Research.* http://doi.org/10.1080/13504622.2017.1392484.

Orr, D. (1992). *Ecological Literacy: Education for a Postmodern World.* Albany, NY: State University of New York Press.

Orr, D. (2004). *Earth in Mind: On Education, Environment, and the Human Prospect.* Washington, DC: Island Press.

Pugh, K.J. (2002). Teaching for transformative experience in science: An investigation of the effectiveness of two instructional elements. *Teachers College Record, 104*(6), 1101–1137.

Rokeach, M. (1973). *The Nature of Human Values.* New York: Free Press.

Rimanoczy, I. (2013). *Big Bang Being: Developing the Sustainability Mindset.* Sheffield, UK: Greenleaf Publishing.

Singleton, J. (2015). Head, heart, hands model for transformative learning: Place as context for changing sustainability values. *The Journal of Sustainability Education.* http://www.jsedimensions.org/wordpress/content/head-heart-and-hands-model-fortransformative-learning-place-as-context-for-changing-sustainability-values_2015_03/

Sipos, Y., Battisti, B., & Grimm, K. (2008). Achieving transformative sustainability learning: Engaging head, hands and heart. *International Journal of Sustainability in Higher Education, 9*, 68–86.

Strong, R., Silver, H. F., & Robinson, A. (1995). What do students want? *Educational Leadership, 53*(1), 8–12.

Schweizer, S., Davis, S., & Thompson, J.L. (2013). Changing the conversation about climate change: A theoretical framework for place-cased climate change engagement. *Environmental Communication,* 7(1), 42–62.

Tooth, R. (2018). *Diverse Pedagogies of Place: Educating Students in and for Local and Global Environments.* London: Routledge.

United Nations Educational, Scientific and Cultural Organization. (2017). *Changing Minds, Not the Climate: The Role of Education.* Paris: UNESCO. https://www.gcedclearinghouse.org/sites/default/files/resources/190248eng.pdf.

Wangaard, D., Elias, M., & Fink, K. (2014). Educating the head, heart and hand for the 21st century. *Southeast Education Network Magazine.* https://www.seenmagazine.us/Articles/Article-Detail/ArticleId/4140/EDUCATING-

Wiek, A., Withycombe, L., & Redman, C.L. (2011). Key competencies in sustainability: A reference framework for academic program development. *Sustainability Science, 6*(2), 203–218.

# 13

# DEVELOPING MINDFULNESS THROUGH INTERNATIONAL IMMERSION

## Transformational learning for a complex world

*Abigail B. Schneider*

### Introduction

As I write the opening to this chapter from my home office in the Western United States, the world looks very different than it did just a few months ago.[1] In China, factories once billowing smoke cease operation as global supply chains grind to a halt. In Italy, an entire country is quarantined in their homes. In the United States, anxious shoppers fight over the last package of toilet paper, and shelves remain bare. The globe plummets into recession. The news reads like a dystopian novel, but the novel coronavirus, and its respiratory disease, COVID-19, is the new reality. Experts warn us that if we continue down our path of rampant consumption, clearcutting the Amazon to create more farmland in hopes of feeding our exploding population, we will only see the rise of more novel pandemics due to our coming into contact with species once unknown to humans.[2] And with global temperatures on the rise, disease will more easily spread. If the recent pictures from space showing China's clearing skies had not already revealed the interconnections between environment, economy and disease, this sobering revelation does: we live in a complex world, and only those equipped to think in systems will have a chance at solving our world's most pressing problems, problems that often disproportionately impact those already marginalized.

Being able to appreciate the complexity of the world and see the interconnections between seemingly disparate systems are hallmarks of sociocognitive mindfulness, a construct associated with cognitive flexibility and coined by Ellen Langer (1989), management professor at Harvard Business School. While I do not claim to be able to solve a pandemic or save the world from a recession, I have strived over the years to create transformative curricula that can help my students start to develop sociocognitive mindfulness as well as a mindset promoting sustainability. The "sustainability mindset" is a concept developed by Isabel Rimanoczy, convener of the United Nations Principles for Responsible Management Education (UN PRME) Working Group on the Sustainability Mindset, who defines it as, "a way of thinking and being that results from a broad understanding of the ecosystem's manifestations, from social sensitivity, as well as an introspective focus on one's personal values and higher self, and finds its expression in actions for the greater good of the whole" (Kassel et al., 2018, p. 7).

DOI: 10.4324/9781003229735-15

In particular, I have developed and taught an interdisciplinary course called *Marketing for Social Change*, which uses a spring break trip to Uganda to teach students about the relationship between economic development, environmental sustainability and human connection in addressing the complex issue of poverty.

Poverty eradication is the first goal listed among the United Nations' Sustainable Development Goals (SDGs), an action agenda adopted by UN Member States to be achieved by the year 2030 in order to establish "peace and prosperity for people and the planet, now and into the future" (https://sustainabledevelopment.un.org/?menu=1300). As the 17th goal highlights, however, all of the goals are interconnected. Thus, in order to solve poverty, students and agents of social change must not only view poverty as an economic issue but also as one intimately tied to environment, gender equality, public health and myriad other challenges. Infusing education with systems thinking and sociocognitive mindfulness can help prepare students to effectively take on these challenges.

In this chapter, I will demonstrate, through students' own words (via their travel journal reflections), how *Marketing for Social Change* helped these emerging change agents to develop sociocognitive mindfulness and a sustainability mindset. I will begin by introducing the concepts of sociocognitive mindfulness and the sustainability mindset in more depth. I will then provide a brief description of my course, *Marketing for Social Change*, which seeks to imbue students with these ways of thinking. Next, I will discuss the learning goals and assignment that led to my students' travel journal reflections, and I will share excerpts from the students' travel journals. I will then provide a short analysis of the travel journal reflections, and I will conclude by providing some suggestions for a way forward.

## Literature review

### *Sociocognitive mindfulness*

The construct of mindfulness is often conflated with a purely Eastern perspective where it is associated with meditation and being aware, in the present moment, without judgment (Kabat-Zinn, 1994). In contrast, the Western mindfulness tradition, specifically the sociocognitive perspective studied by Ellen Langer, focuses on "cognitive flexibility which allows individuals to actively construct novel categories and distinctions" (Pirson, Langer & Zilcha, 2018: 169). Based on an information processing perspective, it is marked by engagement in cognitive exercises such as learning, goal-pursuit and problem-solving (Langer, 1989, 2000; Langer & Moldoveanu, 2000; Pirson et al., 2018). Sociocognitive mindfulness also allows individuals to draw novel-distinctions rather than follow a rigid mindset that relies on prior perspectives regarding distinctions and categories (e.g. routine judgments and stereotyping) without considering present contexts and perspectives (Langer, 2000). As such, those high in sociocognitive mindfulness are more likely to seek novel perspectives, engage in creative activity and generally engage to a greater degree with the present situation (Langer 1989, 2005). In short, "socio-cognitive mindfulness is exhibited by cognitive flexibility that manifests in (a) novelty seeking behaviors such as curiosity and openness, (b) novelty producing behaviors such as creativity and perspective taking, as well as (c) engagement with current activities, people and contexts" (Pirson et al., 2018: 170). As a result, "more mindful individuals [are more likely] to take decisions that are more complex, take more information into account, and serve overall societal well-being better" (Pirson et al., 2018: 179).

Appreciating the complexity grounded in the present context and allowing for the updating of one's mental model of a situation or problem based on that novel perspective is also foundational to the development of a sustainability mindset.

### *Tenets of the sustainability mindset*

In order to effectively address the existential challenges of climate change and sustainability as a whole, Rimanoczy (2013) argues that we must transform the human dimensions of both thinking and being that ultimately result in innovative action for the greater good (Kassel et al., 2018). This thinking and being is comprised of four areas, encompassing the 12 Sustainability Mindset Principles: ecological worldview (ecoliteracy, my contribution), systems perspective (long-term thinking, flow in cycles, both-and thinking, interconnectedness), emotional intelligence (creative innovation, reflection, self-awareness), and spiritual intelligence (purpose, oneness with nature, mindfulness; Rimanoczy, 2021). Particularly relevant to the transformation experienced by the students who engaged in the international immersion trip are systems thinking and the additional dimension of innovative thinking. While innovative thinking is characterized by using novel perspectives to address complex challenges, systems thinking is built on the four dimensions listed above as well as the facets of complexity and cooperation. In this view, "all members of the ecological community" are connected through a "complex network" (Rimanoczy, 2013: 67), and cooperation is critical to their co-evolution. Systems thinking is a holistic perspective focusing on concepts such as "connectedness, relationships, and context" and seeing the world as natural ecologies operate – as "dynamic, large, and complex system[s]" (Rimanoczy, 2013: 67).

From this perspective, the achievement of any Sustainable Development Goal (SDG) fundamentally relies on progress on the other goals. The goals are not independent issues that can be addressed in isolation but rather, they represent myriad manifestations of the status of a larger, complex system. Moreover, as the 17th goal makes clear, success in achieving the SDGs depends on forming relationships, particularly global partnerships, in order to make change at different points in the global network. Thus, any progress at addressing the first goal, eradicating poverty, ultimately depends on coordination among the goals and cooperation across the world, as action at one node in the network has a ripple effect across the globe. Specifically, it is critical to examine the global business landscape and economic network that link the Global North and South in an inextricable but often hidden relationship. Such is the aim of my course, Marketing for Social Change, which I will describe in the following section.

## The course, *Marketing for Social Change*

### *Course description*

The modern global economy is more interconnected than ever before in history. While this integration has led to increased wealth and an enhanced standard of living for some, the current business environment operates on an uneven playing field that has expanded the wealth gap between the rich and the poor (Piketty, 2014; Stiglitz, 2002). While a few hundred million people control the majority of the earth's resources, over 1 billion people on the planet try to survive on less than \$1.90/ day (Collier, 2007; World Bank, 2016; Yunus, 2007). In order to teach students about the inequities inherent in the current global economic system and to invite students to critically reflect on the impact that their own business decisions and consumer behavior have on the planet and people around the globe, I developed the course, *Marketing for Social Change*, which I've taught at Regis University in Denver, Colorado, USA. The course is an elective, taught through the Anderson College of Business[3] as well as Regis College (the liberal arts college), as it integrates the various disciplines with which it is cross-listed (Business/ Marketing, Peace and Justice Studies, Environmental Science, Women's and Gender Studies and the upper-division "integrative" core area of Justice and the Common Good). While the course is intended for third and fourth year students, students of all levels have taken it. The goal of *Marketing for Social Change* is to promote an understanding of the context in which the current business environment and economic system operate and to offer an interdisciplinary (e.g. historical, political, economic, geographical, gendered, cultural, etc.) perspective on the causes and consequences of extreme poverty. Moreover, the course seeks to provide students with a framework for evaluating and developing possible solutions for eradicating extreme poverty from foreign aid and nonprofit work to social enterprise and social marketing.

In the middle of the semester-long course, students travel to Uganda where the professor has personal connections with nonprofit organizations and a social enterprise accelerator that have headquarters in Kampala as well as in Boulder, Colorado, USA, where the professor attended graduate school. Students are in Uganda for about ten days during which time they are able to experience the vibrant Ugandan culture, see the unmatched biodiversity of the country, and meet with conservationists as well as organizations such as the Street Business School by BeadforLife (a nonprofit that supports women in developing their own microbusinesses) and Unreasonable East Africa (an accelerator for social enterprises). Through experiential immersion, the course seeks to give students an appreciation for the interconnectedness of the developed and developing worlds as well as between poverty and other social and environmental issues such as public health, environmental conservation and poaching. Ultimately, the course aims to foster a sense of solidarity with the people and places around the world that are impacted by Western business practices so that future business leaders and policymakers will make decisions in the service of global peace and justice. In summary, the course aims to help students develop sociocognitive mindfulness and a sustainability mindset so that they can think flexibly, manage complexity and act compassionately as effective change agents who will transform the business landscape in service of the public good.

## *Transformation and reflection – Travel Journal assignment*

In order to help students make sense of their experiences and drive personal transformation in terms of thinking, being and doing, one of the focal deliverables assigned in the course is a Travel Journal. Throughout the class, and especially during their travel, students are required to keep a journal, capturing what they did, their thoughts and reflections, and their feelings.

The Travel Journal assignment is divided into the following six sections:

1. An initial reflection written during the first week of class, detailing why the students want to take the class, what they hope to learn, what comes to mind when they think about the continent of Africa, what they imagine extreme poverty looks like, what they believe is the most effective solution to extreme poverty, what they believe they can do

to best help address extreme poverty, what they already know about the continent of Africa and the country of Uganda and how they know those things
2. A second journal entry that is to be written the day before the students leave for Uganda, capturing their thoughts and feelings as they prepare for the journey on which they are about to embark
3. A third journal entry that is to be written while in transit. Students may choose to write multiple entries during different parts of the transit from Denver, Colorado, USA through Amsterdam to Entebbe, Uganda
4. For the fourth installment of the journal, students must write at least one entry per day for each day that they are in Uganda
5. Once students are back in the United States, they are asked to reflect on their thoughts and feelings over the course of the first three days of being home. This reflection constitutes their fifth journal entry
6. At the conclusion of the semester, students write their final journal entry, reflecting on what they learned from the course and travel experience, the impact it had on them, what it taught them about justice and the common good, and how they expect their participation in the class to influence their future. Finally, students are asked to revisit their first journal entry and reflect on what they wrote then given what they know at the end of the course – Have their thoughts, opinions, and/or feelings changed? Have they strengthened? Have they stayed the same? What do they now know about the continent of Africa and the country of Uganda, and how do they know that?

Since I started teaching the course during the spring of 2017, students have shared their impactful and meaningful reflections about their experience in the class, and it is my honor and privilege to share some of those personal stories of transformation below. In particular, I have chosen excerpts that beautifully illustrate the development of sociocognitive mindfulness and a sustainability mindset. In the section that follows, I will introduce each student author, including information about their major, the year that they graduated from Regis University, and their class standing when they took the course and traveled to Uganda. I will then present direct quotes from their Travel Journal assignments that I have deemed demonstrate sociocognitive mindfulness and/or a sustainability mindset. Following the students' quotes, I will present additional direct quotes that the students shared with me in follow-up emails during the Spring of 2020, explaining how the course and travel experience have impacted them since graduating from college. Each student's work and personal information will be presented in their respective sections below.

## Excerpts from students' reflections

**Name:** Brennan
**Major:** Neuroscience with Honors and a minor in Mathematics
**Year graduated from Regis**: 2017
**Year in college when took class and traveled to Uganda:** Fourth Year

> *This city is overflowing with life. Here are some of my favorite sounds … Best of all is when one free spirit (or public nuisance, depending on perspective) plays music loud for all to hear. The song may be*

*ten minutes long, but as we drive by the notes last only for a moment. It's easy to miss, easy to lose that opportunity to share in the experience of others through sound. A good lesson in mindfulness.*

*Thursday night I got mostly packed up but was feeling very down. It was our last night in Uganda. The trip had gone by so quick. I think I was worried that the trip would end, we would go back home, and nothing of substance would have changed. Things would go back to the way they were before. But it's not like I was unhappy with how things were before the trip, just that if nothing changed then what was the point. I had felt such powerful emotions that week, from curiosity to contentedness, depression to exhilaration, fear to tranquility. Uganda had changed who I was as a person, and I* ***really*** *liked that version of myself. I was worried that when I left Uganda I would not be able to bring that home with me.*

*Our last business visit of the day was to Spouts of Water. They make ceramic filters, which are essentially very sophisticated flowerpots placed in a bucket. Kathy, the co-founder, gave a good perspective we hadn't seen yet: how difficult it is to start, grow, and maintain a social enterprise in Uganda as a foreigner. She was honest. Brutally honest. She told us bluntly that Spouts could not have survived without a few bribes. Up until this point I think people had kind of glossed over stuff like this in an attempt to not scare us away … Though the meeting didn't exactly leave a positive taste in my mouth, it was the first time I felt like we had been told the complete truth about what it was like to run a business in Uganda. I'm very grateful she was honest with us …*

*Later that day on Monday, I found a quote while reading Gandhi for another class that really rung true for me about the determination, grace, and passion we had seen in the people we talked to on the trip. It also speaks to both the ability and responsibility we all have to change the world around us. It speaks to challenges on a global, local, and even deeply personal scale. I also find great inspiration in a quote that my step dad uses very frequently. It is the quote I used in my toast on the shore of Lake Victoria, and I think it perfectly captures the personal transformation many of us went through on this trip. Finally, I will put forth a quote from Tolkien that captures the excitement, ambition, and peril of adventure, and a complimentary quote from Socrates that urges us to look deeper at our own lives. Together these quotes remind me to always break out of my comfort zone and go further than I thought I could, but still be mindful and to never lose sight of the bigger picture.*

*That said, how I see my place in the world has changed over the course of this semester. Coming in, I already knew I didn't want to pursue basic neuroscience research. I was looking for a path, a road to follow either in the short term or the long term. I was looking for a sense of direction, and this class has given me that.*

### *Where is he now (Summer 2020)?*

*I am the director of product development at a small (12 person before Corona, 2~3 – person after) web and app development company called Beyond4. We are involved in multiple industries but specialize in restaurant marketing … Of late, our most notable projects have included ensuring that our clients' websites are as accessible as possible to people with disabilities who may be using screen readers and other assistive technology. I have really appreciated the opportunity to work on something like that, as I feel it has positive effects beyond generating revenue.*

*(Magis[4]!)*

## Impact of class overall

*I did return to Uganda the following year and stayed three weeks with an organization called Kin Initiative. They operate in Kampala, Gombe (Wakiso district), and Fort Portal (Kabarole district). Their areas of focus are education and agricultural development. My mission for that trip was very similar to that used during our class visit – to be exposed and listen to the stories, missions, and perspectives of Ugandans to help me become a better advocate for social justice and a better global citizen. As part of this trip I took many photos, some of which are published on my personal website and in social media posts for the organization.*

*I cannot put into words the degree to which I have been shaped by my experiences in your class. I find myself filled with many different emotions as I consider the level of impact it has had on my job, my worldview, and my personal philosophy. Because of your class, I am a more centered, driven, and effective individual. Your class helped me develop a vision and sense of purpose that I will rely on throughout my entire career and life.*

## Impact on career

*Our company is split between the US and Peru. When I started here, this wasn't really known to me, but it quickly became one of my favorite things about the job. Our sales, design, and project management were all based here in Austin while our development was based in Lima – each office being led by one of the two co-founders. I feel that your class really prepared me to function well in and draw deeper value from this cross-cultural business environment … in September 2019 and January 2020 I had the privilege to travel to Lima and work in person with my colleagues. Both of these experiences were very valuable and were meaningful continuations of my development as a global citizen, which I believe was really accelerated by your class. I have learned a tremendous amount about the operational benefits, challenges, and greater personal value of working in a multicultural and multilingual organization.*

*I only recently realized (after getting your email) the insane level of continuity between your class and where I am now. Your class primed me to think of how commerce, especially marketing, can be used for the common good. I believe this is incredibly powerful, as we can take an existing system which has a mixed track record when it comes to social justice, and repurpose it as a deliberate tool for the improvement of our society.*

*I have started to develop a vision of business as a vehicle for maximizing the human potential of all parties involved: the employees, the managers, and the communities in which they operate. As my career continues, I would really love to make justice-oriented, cross-cultural leadership a central tendency. I guess that is what social entrepreneurship is all about!*

**Name:** Bailey Gent
**Major:** Business Administration and Peace & Justice Studies
**Year graduated from Regis:** 2019
**Year in college when took class and traveled to Uganda:** Second Year

*My impression of how to do global work has changed in some ways from the things I learned while in Uganda and while traveling to and from Uganda. I feel differently about some of the ways that I think we as westerners work in these spaces. A lot of those feelings are big feelings that*

*I am still processing and sifting through. In a lot of ways, I have always wanted to work globally, have worked globally, and have encouraged global work in nonprofits and social enterprise. While I went into this trip feeling like I would be encouraged by the benefits of those models, I also saw a lot that discouraged me in what was happening too. Seeing and feeling these things pushed me towards believing some different things about the work that is being done.*

*This class related in some points as well to the concept of justice and the common good. The way that I interpreted justice and the common good relating most to the class was in the spaces when we talked about our place in the world and if it is our responsibility/ if we have the right to intervene in these places putting our own thoughts and beliefs into action. Those conversations are incredibly good for my brain because being challenged by what seems logical to me is good for rounding out my worldview in life. I have for many years believed that it really makes sense for those of us who have plenty to give to others. For those who are educated to work to create sustainable solutions for pressing issues in the world today. However, I am also painfully aware that there can be harm in even the most well-intentioned versions of doing those things. However, it was on this trip that I truly did question if this work is good. So that is an impact that I am still toying with I suppose and also a justice and the common good thought as well. However I do still feel a bend towards believing that we do all have a responsibility to care for one another. I have to believe that, that is how I live my life and the touchstone by which I do all of the things I do in my life. In my heart and in my mind though I am left with a lot of questions and not a lot of answers.*

### *Where is she now (Summer 2020)?*

*Working as the CareerWise Colorado Training Specialist designing career and management training for a modern youth apprenticeship program.* https://www.careerwisecolorado.org/.

### *Impact of class overall*

*I would love to go back to Uganda and hope to with my friend's nonprofit in the future. At this time I have not returned but I did go back to Africa shortly after this trip and was in West Africa at that time. My time in Uganda was definitely on my mind while I was in Ghana and influenced some of my thinking on the work I did on that trip. After graduation, I also worked for an international development organization and used my knowledge from my time in Uganda in that capacity as well.*

### *Impact on career*

*I think that it did encourage me to want to work in education in a way I did not expect beforehand. In the future, I hope to continue working in women's education globally.*

**Name:** Lindsey Furton
**Major:** Mathematics, minor in biology
**Year graduated from Regis**: 2018
**Year in college when took class and traveled to Uganda:** Third Year

*I think that putting oneself out of the comfort zone is an essential part of traveling to a new place, and it is what fosters growth, strength, and knowledge.*

*I was so interested to see and hear about the healthcare that my group received [in Uganda] and really felt like I wanted to do something to improve that quality of care. This was when I realized that I definitely should do public health work internationally. I also realized that my perception of healthcare in Uganda was very off until I actually visited the country, and I still would need to spend more time to learn truly about what it is like. Secondly, I realized how culture affects things, as I learned that it is rude to question the doctor. These are things that I will have to keep in mind when I do public health work internationally.*

*We had the opportunity to meet with some of the Unreasonable East Africa businesses today. My favorite one was WANA Energy Solutions because I love what they are doing. In our meeting with the founder, he informed us that ninety percent of the population in Uganda uses firewood for cooking and this decreases the air quality, and in thousands causes chronic bronchitis or even death. This happened to his mother, and it motivated him to start this company. He sells propane and stoves for cooking, and they work to make it affordable for people to use … I also really like that he was solving a health problem in a different way than I had ever thought. Before going on this trip, I thought that the diseases prevention and treatment in Uganda could be accomplished simply through providing vaccinations and essential medications. I never realized that big public issues could be solved by something so simple as changing method of cooking, and it's not something I would have ever thought about until visiting the country. This is yet another thing that I can keep in mind in my future career.*

*I learned that I for sure want to do public health work internationally, and that in order to do it effectively, I have to learn from the people and the place itself, keep in mind cultural differences, and think outside of [the] box of health care solutions that I am used to.*

*I feel like such an improved person coming back from this trip in so many ways that I can't even describe.*

*I think that having the ability to question the environmental, political, cultural, historical, etc. effects on public health issues, and having that experience will help me greatly in my public health career. I will probably be looking at a lot of statistics, but knowing the importance of learning from people and looking into all of these factors behind the numbers will be of key importance to success in a public health career.*

*From BeadforLife, I learned that there are so many factors that go into the improvement of health in a population. It's not only important to supply the right kinds of vaccines and medicines, but it is also important that families have access to them. BeadforLife taught me that education could be the first step in helping families to learn about healthcare and help them to make the money to be able to purchase it.*

*Looking back on my first journal entry, it's crazy to see how much I have learned throughout the semester. In my first journal entry, I wrote, "I think that the most effective solution to poverty is education," and I proceeded to explain why, and used BeadforLife as an example. I remember writing this, and I remember how simple I thought the answer was. Although education is a very big part of the solution to ending poverty and other social issues in Uganda, and even though education through the BeadforLife program has rippling positive effects, there are so many other things that prevent Uganda from reaching its full potential. Like I discussed earlier, there are geographical, historical, health, cultural, and other barriers that would be holding Uganda back even if education were improved in the country.*

*I propose a different solution that isn't simple, but it truly would make the biggest difference in relieving Uganda and other countries from poverty. I realized that if aid could be given with more care and understanding, so that the money helps rather than hurts, and is not put towards causes in the self-interest of the Global North, then that would be an improvement. I think that it could be helpful to open up trade barriers and allow Ugandan imports in the United States, which would be great for Uganda's GDP. I also think that the Global North needs to halt in its resource extraction from countries like Uganda, because it can often make the land uninhabitable and it's unfair, as the profits don't go to the citizens most of the time. In the neoliberal world that we currently live in, the gap between rich and poor, Global North and South is widening, and it gives the greater groups a sense of power over not only the material things of our planet, but also a sense of power over others, and it is because of this that their harsh actions towards other parts of the world are justified and continued even after seeing devastating effects. I think that in order for big, positive changes to happen, there must be an evening out of power between the Global North and the Global South. I think with that, the North would be more responsible and caring in its actions toward the South, whether through aid, trade barriers, the accumulation of resources, etc.; the South could hold them accountable; and countries like Uganda would be able to enter the global market without being held down by neoliberal forces.*

### *Where is she now (Summer 2020)?*

*I am a student in the Master of Public Health in Epidemiology and Biostatistics program at Boston University.*

### *Impact of class*

*In traveling to Uganda through this class, I had the opportunity to meet with different organizations that do amazing work promoting health in Uganda. I saw how these organizations used their deep understanding of the culture, history, and environment of Uganda to find innovative solutions to public health issues. This influenced my decision to pursue a graduate degree in public health. I hope that I can similarly grasp the big picture of global health issues and use my degree to help make a difference.*

## Analysis and discussion

As the above reflections demonstrate, the course, *Marketing for Social Change*, helped the students to develop sociocognitive mindfulness and the ability to see the bigger picture and nuances of issues. They advanced their cultural understanding, and they developed an appreciation for differences in traditions and perspectives. The students were open-minded, allowing the present context (their experiences in Uganda) to shift their mindsets, and they were willing to revise their opinions about the most effective solutions to solving extreme poverty as well as their ideas about what their role in that effort should be. While the assigned readings in the course[5] certainly helped to contextualize the travel component and give students an intellectual basis for understanding extreme poverty and related issues, it was the travel component that allowed these students to adjust their own perspectives in relation to the academic knowledge they were acquiring.

Both Brennan and Lindsey discussed the importance of moving beyond their comfort zones, and as Lindsey noted, "traveling to a new place … is what fosters growth, strength, and knowledge." Such a perspective is echoed by systems thinker Peter Senge in his book, *The Necessary Revolution* (2008). When discussing the week-long trips to Brazil taken by members of the Global Sustainable Food Lab, an initiative that seeks to make sustainable food systems mainstream by bringing together multi-sector stakeholders such as Unilever, General Mills, Green Mountain Coffee, and the World Wildlife Fund, Senge (2010) writes:

> Learning journeys are expeditions taken in search of a new understanding of an issue or a set of issues. Learning journeys mean leaving the familiar behind and going to see unfamiliar aspects of a system firsthand … Complex issues often require travel by groups (sometimes to other regions or even other countries) who are seeking to build a collective understanding of a system or a part of a system … The key is moving outside of the familiar … We all know some parts of the system we belong to firsthand, but we know many other facets only through books and articles, or through reports and figures … learning journeys not only take people physically to places they have not been, but open awareness beyond the intellect. "Going there" means seeing a different place firsthand, talking with people we usually do not talk with, seeing how people live, work, play, raise their children, and help those who are sick: Only then do we start to sense, rather than just understand intellectually, a larger system. Especially important is to use the direct experience to consider views very different from our own, views held by people whose historical experiences of the system we belong to may differ.
>
> *(p. 260)*

By traveling to Uganda, students were able to interact with those whose experiences and histories different greatly from their own, and they were able to develop an appreciation for the complexities of the issues they desired to address. What students once thought could be viable solutions to poverty turned out to be only a single piece of a much larger puzzle of interconnected issues. Through the experience of travel, the bigger picture of interrelated issues and the connection between the Global North and South started to become more visible to the students.

Moreover, students reexamined the role they believed was appropriate for them to play in addressing those issues. As Bailey noted, she wants to work in service of the marginalized, but she wants to do so in a way that does not reinforce colonial paradigms. The students were and continue to be sensitive to the power dynamics endemic in the global economic system, and they want to ensure that their work transcends systemic inequalities. They have genuine respect for those they serve, and while in Uganda, their hearts were deeply touched. According to Senge (2010), developing the heart, as well as the mind, is a central feature of learning journeys.

Such pivotal experiences can also be critical to the development of a sustainability mindset. In her research, Rimanoczy (2013) found that many business leaders who worked toward sustainability were inspired to do so after experiencing a life-changing event (e.g. travel, encounters with people whose experiences differed from their own, etc.). By engaging in an experience or going through a major life event, the participants in Rimanoczy's study underwent great transformation.

Likewise, the students who took *Marketing for Social Change* were aware of their own personal transformations and the impact that the class and experience had on their career aspirations and who they want to be as people in the world, and they continue to reflect on that impact.

Such reflection is a critical component for realizing the benefits of travel learning. As Senge (2010) writes:

> With enough time for reflection and conversation, learning journeys can start to reveal a third possibility … conversations that open hearts as well as minds. Gradually, relationships based in genuine respect and empathetic listening can start to grow … Learning journeys, at their essence, are about "sensing" (or opening awareness to the present moment), but the goal is not simply awareness for its own sake, or to only deepen relationships between key players, but doing this in the context of compelling issues about which people care deeply. The team members must ultimately be prepared to reexamine their own beliefs and tried-and-true approaches in the service of genuine change.
>
> *(pp. 261–262)*

While *Marketing for Social Change* certainly attracts aspiring change-makers, it also risks leaving students wondering how "genuine change" is possible and how they can be part of it. Once students realize how big and complex issues such as poverty truly are, it is easy to become faced with the challenge of "analysis-paralysis" (Pirson 2017: 157). Those who have witnessed the system cannot view poverty as a one-dimensional (e.g. economic) issue with a single solution, and it can be easier to critique the failures of prior attempts at solutions than to imagine new ones. Moreover, some wonder whether, morally, they should even be part of the change. Like Bailey, they are sensitive about replicating colonial tropes and question whether those in poverty would be better off without help from the West. Ultimately, however, there is no choice. The very history of colonialism has led, through our global supply chains and economic system, to an inextricable link between the Global North and South.

As Naomi Klein (2000) writes in her book, *No Logo*:

> It would be naive to believe that Western consumers haven't profited from these global divisions since the earliest days of colonialism. The Third World, as they say, has always existed for the comfort of the First. What is a relatively new development, however, is the amount of investigative interest there seems to be in the unbranded points of origin of brand-name goods. The travels of Nike sneakers have been traced back to the abusive sweatshops of Vietnam, Barbie's little outfits back to the child laborers of Sumatra, Starbucks' lattes to the sun-scorched coffee fields of Guatemala, and Shell's oil back to the polluted and impoverished villages of the Niger Delta.
>
> *(p. xviii)*

In other words, it is not a question of whether we, as Westerners, should get involved with efforts to eradicate poverty but rather how we need to change the way in which we are already involved. As Lindsey describes, once we understand our position in the larger system and our role in perpetuating the injustices we seek to address, we can change our consumption behavior and business practices that impact those at other places in the system.

Such is the goal of *Marketing for Social Change*. Unlike traditional service learning classes that engage in direct service at the place where symptoms of systemic ills are felt, *Marketing for Social*

*Change* seeks to cultivate an awareness of the larger system and an appreciation for the complexities of and interconnections between issues such a poverty, environmental sustainability, public health, gender equality, education and peace and justice among others represented by the SDGs. With a focus on listening and learning, students are able to develop authentic relationships and the empathy and compassion that emerge when solidarity is nurtured. And it is the hope that such solidarity will motivate responsible consumption and ethical business decision-making, much as it has for Brennan. As Brennan noted, his experience in the course – as well as his subsequent travel to Uganda, which focused on listening and relationship building – has had an important impact on how he approaches international business in his current job.

Of course, it may not be possible, even within the framework of the current course, to escape the power dynamic resulting from the history of colonialism. Who has the privilege of embarking on the learning journey? Who has the freedom of global movement? Who undergoes transformation as a result of the class? What is the environmental impact of traveling around the world? These challenges have not only raised more questions than answers for the students but also for me as the professor. Nevertheless, like Bailey, I do believe that there exists a universal responsibility to care for others and our planet, and I believe that transformational learning experiences that help foster sociocognitive mindfulness and an ability to see the complex systems of which we are part so that we can better understand how our actions affect others across the globe is one way to at least start building a collective consciousness that can help address at least some of the SDGs.

## Conclusion

As the world teeters on the precipice wrought by the current global health crisis, we are reminded that COVID-19 is not functioning in isolation but is also related to the mounting economic crisis as well as the crises we have long been facing: climate change, environmental degradation, overconsumption. If we are to try to address any of these issues, we must do so with the understanding that their nature is a truly systemic one and one that must be addressed through cooperation from stakeholders and citizens around the globe.

By engaging in international immersion, students (as well as business leaders) are led outside of their comfort zones and are exposed to people, places and ideas that can help foster innovative solutions and the sense of solidarity that is necessary to enact them. By developing sociocognitive mindfulness and a sustainability mindset, those who have experienced international immersion are able to engage with the complexities of the challenges related to the SDGs, and they are able to do so in a way that acknowledges their place in the system, that is ethical, and that truly promotes well-being on earth. Of course, challenging one's assumptions and worldviews and leaning in to the messiness of reality can be unsettling, but for those who have developed sociocognitive mindfulness, sitting with the discomfort that this awareness brings is but a crucial step along the transformative journey on which they are determined to embark.

## Notes

1 The opening to this chapter was written in mid-March of 2020. The trajectory of the coronavirus pandemic around the globe and the subsequent economic downturn in the United States has since taken many turns and may change again by the time of publication or reading.

2 And in the United States, a reliance on factory farming (a major contributor to climate change) is likewise increasing the odds of another pandemic due to the overuse of antibiotics when keeping livestock in close quarters. Ironically, the US government refused to shut down meatpacking plants after they became hotspots of COVID-19 outbreaks and workers (many of whom are immigrants and refugees) had their health put at risk.
3 In 2020, the college was restructured and rebranded as the Anderson College of Business and Computing.
4 "Magis" is a term central to the Jesuit Catholic tradition, meaning "the more universal good" (Geger, 2012).
5 Readings included works from authors such as Abhijit Banerjee and Esther Duflo, Jared Diamond, Nicholas Kristof and Sheryl WuDunn, Wangari Maathai, Dambisa Moyo, Jeffrey Sachs, Joseph Stiglitz and Muhammad Yunus, etc.

## References

Collier, P. (2007). *Bottom Billion: Why the Poorest Countries are Failing and What Can Be Done About It*. New York: Oxford University Press.

Geger, Fr.B.T.S.J. (2012). What magis really means and why it matters. *Jesuit Higher Education: A Journal, 1*(2), 16.

Kabat-Zinn, J. (1994). *Wherever You Go, There You Are: Mindfulness Meditation in Everyday Life*. New York: Hyperion.

Kassel, K., Rimanoczy, I., & Mitchell, S.F. (2018). A sustainability mindset model for management education. In K. Kassel & I. Rimanoczy (Eds.), *Developing a Sustainability Mindset in Management Education* (pp. 3–37). New York: Routledge.

Klein, N. (2000). *No Logo*. London: Flamingo.

Langer, E.J. (1989). *Mindfulness*. Reading, MA: Addison-Wesley Pub. Co.

Langer, E.J. (2000). Mindful learning. *Current Directions in Psychological Science, 9*(6), 220–223.

Langer, E.J. (2005). *On Becoming an Artist: Reinventing Yourself Through Mindful Creativity*. New York: Ballantine Books.

Langer, E.J., & Moldoveanu, M. (2000). The construct of mindfulness. *Journal of Social Issues, 56*, 129–139.

Piketty, T. (2014). *Capital in the Twenty-first Century*. Cambridge, MA: The Belknap Press of Harvard University Press.

Pirson, M. (2017). Social entrepreneurship. In Congress, E.P, Luks, A., & Petit, F. (Eds.). *Nonprofit Management: A Social Justice Approach* (pp. 153–159). New York: Springer.

Pirson, M.A., Langer, E., & Zilcha, S. (2018). Enabling a socio-cognitive perspective of mindfulness: The development and validation of the Langer Mindfulness Scale. *Journal of Adult Development, 25*(3), 168–185.

Rimanoczy, I. (2013). *Big Bang Being: Developing the Sustainability Mindset*. Sheffield, UK: Greenleaf Publishing.

Rimanoczy, I. (2021). *The Sustainability Mindset Principles: A Guide to Develop a Mindset for a Better World*. New York: Routledge.

Senge, P., Smith, B., Kruschwitz, N., Laur, J., & Schley, S. (2010). *The Necessary Revolution: How Individuals and Organizations Are Working Together to Create a Sustainable World*. New York: Broadway Books.

Stiglitz, J. E. (2002). *Globalization and Its Discontents*. New York: Norton.

World Bank (2016). Understanding poverty. https://worldbank.org/en/topic/poverty.

Yunus M., (2007). *Creating a World Without Poverty: Social Business and the Future of Capitalism*. New York: Public Affairs.

# 14

# BRINGING THE CONCEPT OF POVERTY INTO CLOSER FOCUS USING ONLINE GAMES

*Isabel Rodríguez-Tejedo*

## Introduction

This chapter will describe an unusual way of teaching a class about poverty, part of the course on "Global political economy," a mandatory undergraduate class taught in the School of Economics and Business at the University of Navarra (Pamplona, Spain). To facilitate a mindset shift, we sought to engage the minds, but also the hearts of the students. The activities were specifically selected to complement traditional theory so students would think and learn as well as reflect and feel. The combination was hoped to encourage them to revisit their own assumptions and even move them to action.

The course is divided in two groups that meet twice a week during the spring semester. Typically, about 170 students register for the course every year. The student body is made up of students from the 2nd year of the Business degree (about 35%), 2nd year of the Economics degree (10%), 3rd year of the Law + Economics (5%) and Law + Business (15%) degrees, 4th year of the Economics + Governance + Leadership degree (20%) and exchange students (15%). About half of the students are Spanish, and half international, and there is a roughly equal distribution of men and women.

Global political economy (GPE) uses some of the models and concepts covered in the introductory courses of economic theory, such as imperfect competition, market imperfections, comparative advantage, etc., to explore relevant issues in the global environment. GPE covers topics such as globalization, trade, poverty and development, financial crises, energy and sustainability, gender economics … some of which may be switched in or out depending on time and what is most relevant in the current global environment. The course aims to develop competences such as "apply concepts, theories and models necessary to form opinions," "gather and interpret data to make judgments," "interpret graphical data," "develop critical and independent thought," "defend personal ideas on economic issues" and "understand the influence of the economic context on business activities." An overarching objective of the course is to reflect on values, mental models, assumptions and paradigms, to facilitate a mindset shift that may carry over into the students' future professional practice.

DOI: 10.4324/9781003229735-16

The part of the course where the activity described in this chapter is included deals with poverty and development, one of the topics that is always part of the course. The learning goals for the poverty section include getting a solid knowledge of empirical findings, metrics and policy approximations. Although not a direct learning goal, this chapter also aims at spurring individual growth that may inspire future graduates to include sustainability concerns in their professional practice.

## Motivation

Traditionally I employed lectures, videos and readings to cover the material. Students learned the facts and theories about poverty and development well enough with this method, but class discussion and student feedback signaled strong disconnection from the topic and its more human aspects. This was obviously not true for all students, as some were powerfully motivated by the material itself, but in general students approached the study of poverty as they did the comparative advantage theory of trade: something to learn and do well on, but not something that particularly affected them personally. As Rimanoczy (2021) explains, it is important to understand the external factors, but also the individual beliefs, values, assumptions, mental processes and paradigms we all hold. This rang true in the study of poverty in Global Political Economy, and acquiring theoretical knowledge fell short of what I hoped students would achieve after finishing the course. Poverty is at the core of the United Nation's Sustainability Goals, which are a call to individuals and societies to act to achieve a better world for all.

To complement the formal training already in place I set out to cover the topic in a journey of sorts, where the class works together to reflect on what causes poverty, how it is experienced individually and how it can be measured so that appropriate policies may be implemented. We also decided to introduce experiential learning via online games. Because of the large class size, we decided to take advantage of simple technological resources, both to create a repository of student feedback and to create engagement.

Over the last two decades higher education has made tremendous strides in the use of technology. For example, Educase's survey of 2018 found that a majority of students report their teachers use technology to enhance their courses. However, despite its potential, sometimes our use of technology is limited to complementing the traditional lecture, where the student is a passive recipient. Some easy-to-use tools, such as Kahoots, online discussion boards, Q&A + short reflections via Google forms and online games can start to change this practice.

Games in particular present the opportunity to change this dynamic in a low-risk way: they don't require a lot of class time or prep work and students are usually amenable to the experience, so they don't add much to the rigors of a long class. These "internet-based learning environments where decisions are made within a complex and dynamic setting, and where students experience real-time information and feedback" (Lovelace, Eggers & Dyck, 2016) shift the focus from the lecturer to the student and create opportunities for a mind shift that are not easy to achieve via lectures, readings or other traditional methods. They are not too time consuming and are easy to implement, making them an interesting addition to a journey-like experience that tries to join theory with more student-centered activities. The blend of games and traditional lectures is meant to improve learning. Bonde, Makransky, Wandall, Larsen, Morsing, Jarmer and Sommer (2014) found the combination of the two methodologies to be superior to either method alone.

There is a significant amount of research highlighting the potential benefits of including game-based learning in higher education. Farrell (2005) found that students considered games to be more effective than textbooks and cases. Bonde et al. (2014) and Day-Black, Merrill, Konzelman, Williams and Hart (2015) both found students were more interested and did better if games were included. Allcoat and von Mühlenen (2018) concluded that students using virtual reality show increased positive emotions and higher engagement, compared to passive video watching and textbook based learning. Crocco, Offenholley and Hernandez (2016) even found evidence suggesting that game-based learning can increase deep learning,[1] probably by increasing enjoyment. A more complete revision of the literature may be found in Subhash and Cudney (2018).

## Description of the activity

The activity was designed to turn the poverty and development topic from "a set of theories and facts to be studied for a test" to a cooperative learning experience that might facilitate a shift in the students' perspective.

To introduce the topic, the instructor shared with the class the results of a survey administered earlier in the semester where the students had answered the question "in your opinion, what is the most important Sustainable Development Goal?" About 25% of the students answered "SDG1. No poverty," 15% selected "SDG2. Zero hunger," roughly 5% chose "SDG10. Reduced inequality" and the rest of the answers were distributed among the remaining 14 goals. Almost half the class thought this was the single most important issue! This was a powerful starting point, and also allowed the class to re-connect with the learning expectations they had at the beginning of the semester.

Students were then invited to come up with their own answers about questions such as "what is poverty?," "what causes it?," "what needs to be done to get out of poverty?" and then discussed them in small groups. Their answers were collected via a Google form. Virtually all of the answers described poverty as a "lack of." Most of the definitions concentrated on lack of money or resources (about 90% of the answers). Some went slightly deeper and discussed issues relative to lack of access to services or "a decent life," and some came close to definitions of relative poverty (where a household is poor if its income falls below a percent of the median household, instead of using a fixed amount of money). Only two answers went beyond the idea of poverty being an absence of things/ services:[2]

> *A multifaceted issue, characterized by many levels of insufficiency. Extends beyond a lack of material wealth.*
>
> *I would say that poverty is a state of mind.*

A summary of the results was presented to the students. This proved to be a good starting point to describe the complexity of poverty, and also set a tone for discussion and conversation that can be carried over for some time. The class then moved on to more traditional discussion of the facts, measures and definitions of poverty, highlighting its multidimensionality and especially how poverty is not necessarily restricted to money. Through this part of the topic, traditional lectures were complemented with graphs, videos, small group discussion, kahoots and the reading and student presentation of current articles and editorials, to balance the pedagogical approach and increase student engagement.

Students then took the Gapminder test (Gapminder, 2020; see <https://www.gapminder.org/test/>), a 13-question online quiz that tests knowledge of basic realities of global development. Taking the test in class allowed us to focus on the message that if we don't know what the facts are, all we have are our (biased) opinions. Good policy and good decision making in businesses needs to be based on factual evidence, and we cannot seek to learn something if we believe we already have that knowledge. Students were positively shocked at how few correct answers the class got. As expected, the shock served as a powerful motivator. In the students' words:

> *with the gap minder I understood that I didn't know as much about poverty as I thought.*
>
> *the whole class filled in the Gapminder Test and we saw how little we actually know about poverty. We kind of live out[side] of the reality that it represents.*
>
> *I could see that my colleagues … were definitely impressed by the results of the questions of the test.*

It proved useful to pair the test with an excerpt of the first few minutes in Hans Rosling's video "The best stats you´ve ever seen" (The best stats you´ve ever seen, 2020; see https://youtu.be/usdJgEwMinM). The video served to highlight that we are not alone in not knowing, but it is our job to do better! This message was repeated a few times while learning about the facts of poverty and development – a tedious segment that often is hard for the students. As a closing activity for the topic, students were asked to play two free online games, one focusing on the developing world and another in the developed world. This allowed students to see the differences between poverty in both cases, but also to realize that poverty is a very close-to-home reality. "Ayiti, the Cost of Life" (Ayiti, the Cost of Life, 2020; seehttps://ayiti.globalkids.org/game/) and "Spent" (SPENT, 2020; see http://playspent.org/).

"Ayiti, the Cost of Life" simulates the life of a five-person family living in poverty in rural Haiti. The player needs to make decisions about work, education, health and leisure for the three kids, mom and dad during four years so that the family remains healthy and happy, while also getting an education in the form of "diplomas." Each character in the game has stats for health, happiness and education and they share the family´s resources, a sum of money that can be spent on consumer goods (such as for example a school uniform, shoes or a bike), medicines, tuition and so on.

Keeping with the rhythms of a rural community, the years are divided in four seasons. The player must choose each season how to spend the money and assign each family member to a role for the season (working in the family farm or in work-for-hire, resting at home, attending the local school, volunteering). Random events (such as a storm, disease or money received from a cousin abroad) can alter the family's fortune, and the task assigned for the characters can sometimes be modulated by choosing to "take it easy" or "work hard." The game provides notifications about progress and character stats updates. The visuals are simple, but appealing, and keep the player's interest.

"SPENT" was created by the Urban Ministries of Durham, to create awareness about poverty and homelessness in the US. The game starts out with an interpellation "Urban Ministries of Durham serves over 6,000 people every year. But you'd never need help, right? Prove it – Accept the challenge." As the player enters the game, he gets the dim starting point: "Your savings are gone, you've lost your house and you're down to your last 1000$. Can you make it through the month?" The game begins at day one of the month, when the player gets to choose among three low-paying jobs and then is walked through some basic selections (health care, housing) which highlight trade-offs between affordability and convenience. As days go by, the player is presented with a series of hard choices that closely resemble real life situations for those living in or close to poverty in the US. Each choice has an impact on the amount of money left, but also on things like whether her child can attend a birthday party, how much debt she goes into, etc. Although the game is entirely text-based, its use of realistic choices and impactful wording make it more realistic. One particularly interesting feature is that, as the player chooses, the game sometimes provides facts related to the choice and its consequences. For example, when the player keeps a sick child at home while going to work the game reports that "A recent study found that parents in Baltimore, Maryland, were more likely to leave a sick child home alone all day than parents in Honduras," and after having a hard time managing grocery shopping under a tight budget the game signals. "Almost 15 percent of American households had a hard time getting enough to eat at some point during 2012." The game ends when the player runs out of money or the month is over.

The original lesson plan included a short recap of the student answers to the questions of "what causes poverty in the developing and developed world?" and "what would be needed to get out of poverty?," followed by individual playing of each game during class time and a group discussion. Due to the home lockdowns brought about by COVID-19, only one group of students got to play Ayiti in class, and everyone had to play SPENT from their homes. Although unplanned, this did provide an unexpected insight: quite a lot is lost if the discussion and in-game and post-game interactions disappear.

Because the games are online it is possible to ask students to play them outside the class. However, based on the very limited experience we had in 2020, my intuition is that devoting synchronous class time to the playing of the games is highly beneficial. Having all students playing at the same time leads to more insightful discussions and elicits a more personal response as students tell each other about their difficulties and feelings. It also makes it easier for the instructor to connect course content to the game. As the class bounces ideas off one another, the instructor has the chance to tie them back to concepts in the course, write interesting issues on the board for further discussion and in general walk around the room sharing and facilitating conversations. If playing the games in the same physical room is impossible, one may consider splitting the larger online group into smaller ones so students can have a shared experience via Zoom, Google meets or a similar platform. A shared virtual board, be it Padlet or even Google docs, could also help in providing a common repository similar to what would be created in the classroom.

Right after students played the game, they were asked to answer a few questions in Google forms about how they chose to play the games and what their impressions were. Although immediate responses varied greatly, many answers revolved around the idea that playing the games had brought a new perspective on what living in poverty looks like. In many cases, students mentioned that it had made them rethink some previously well-set impressions. Here are some immediate impressions students shared about the games:[3]

Some selected feedback from "Ayiti: the Cost of Life":

> *it made me realize so many things and "reorder" the way I put importance to some things in life.*
>
> *This game was very beneficial. Even though it's just a game it gave me an idea of how hard life can be on some people and the challenges poverty brings. This is real, people die every single day in situations like this.*

*I realized life needs to be balanced. I thought work was always the answer but I don't think the same anymore.*

*It lets you actually understand the fact that getting out of poverty is a reality much harder and complicated than we usually tend to imagine. There is no apparent right path and even the strategy you think might be the most beneficial for everyone could to not work in the end.*

Some selected feedback from "SPENT":

*Helps me to put in the shoes of those people and change how I see the world around me. I feel like I need to do something.*

*This game has been beneficial in order to have a wider perspective on this people who try hard to make their livings.*

*It has made me think hard about how there are many people who are close to us and that have to deal with this kind of situations daily. I see things from another perspective now.*

*I could not avoid difficult situations. It has opened my eyes to other people's lives and how I need to make a difference.*

*This game made me realize that being poor in a developed country is also very hard. And more factors come into play, such as your children and his care, you health, your stress levels. It was all a day to day decision making process. It really opens your mind when you live pay check to pay check, you put things into perspective.*

*It has changed my mind about how difficult it is to make it until the end of the month. I see poverty in a developed world different and I want to do something to create change.*

*This help me realize that is easy to get into poverty, but not to go out.*

A few days after finishing up the chapter in poverty, students were asked to complete a personal graded assignment. The assignment asked students to (1) answer some "theoretical" questions about poverty and development, and (2) find as many of the theoretical concepts from the course in the games as possible. In the same assignment (although it was explicitly explained that this part was not graded), students were asked to reflect on whether (and how) their personal views had changed after the chapter, and in particular after playing the games.

In terms of getting students to express their thoughts, this format proved superior to the online surveys conducted immediately after playing the games. It did require more work (setting up and going over the graded part of the assignment), but provided better results. Even though it was not graded, a larger number of students wrote down full reflections and most of those showed a depth that was not present in the reflections collected via online surveys. It is impossible to know why, maybe the higher rate of answers was due to the fact that they didn't have to actively open another document, or because they felt compelled to "finish" the questions.

The following graph on Figure 14.1 presents a rough distribution of the students' reflections.

The results are encouraging: in over 75 percent of the answers the dominant feeling was that the games did have the desired impact. It is worth noting, however, that about five percent

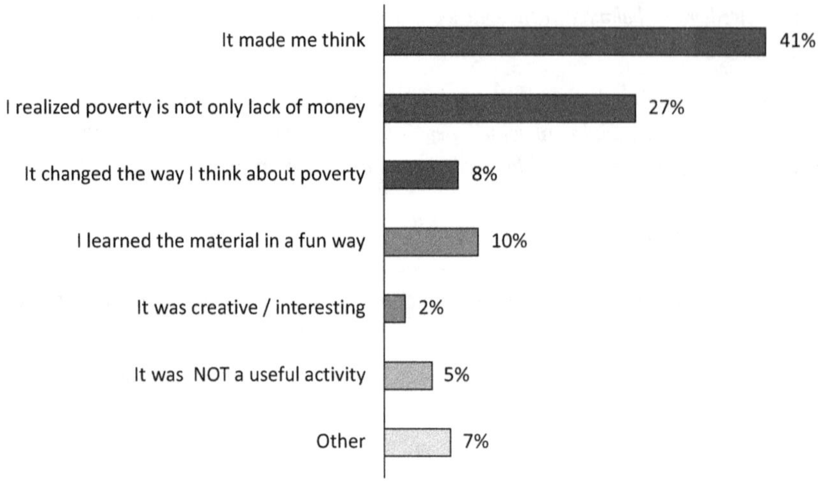


**FIGURE 14.1** Students' opinions about the games

of the students actively reported finding the games not useful. Some students suggested the game didn't add much to their understanding of poverty or thought the games made situations seem unnecessarily hard:[4]

> *My point of view has not changed, I was already aware of the struggle and hardship the people with scarce resources are going through every day.*
>
> *Sometimes it seems that it is always trying to create the worst scenario and making it too difficult.*

To get a better understanding of the impact (if any) of the games and the overall learning experience in the poverty & development topic, I asked some students to recount their experiences in a more free-style format. The following section of this chapter is devoted to some of these first-person stories (names are omitted to preserve anonymity). To include more students' reflections, these are passages instead of full texts.

## Student reflections

### *Reflection #1 (student, University of Navarra, Spain)*

> *I have learned useful tools to understand poverty as it really is, multidimensional. Such complex problems are sometimes approached from reductionist perspectives, (in many occasions, we, the economists, are responsible for creating unrealistic models of reality) and elaborated by people far removed from the subject they are studying. This is not a criticism of the academic world because it is not necessary to be "poor" to study poverty, but it is necessary to get a little out[side] of oneself. Analysis and study is a process not only of observation but also of transformation. With this subject I have realized that learning has to be two-way or it is not learning. That is to say, if I try to understand poverty but I am not transformed from that knowledge, then allow me to question*

*the teaching process. Reading data does not make you learn, but connecting them with yourself, with the world and with your surroundings is what transforms you and that's exactly what this approach encourages.*

*In relation to this, I would like to say that knowing the data is important, but comparing it with previous mental preconceptions was truly revealing. The destruction of biases and the process of cleaning the gaze to later offer a "pure" analysis. I think it should be done in all subjects. Confronting myself with my previous ideas allowed me to open my eyes to new dimensions of poverty.*

*Poverty greatly limits the ability to act in the face of any unforeseen event. These small but fundamental details to understand poverty are those that I have been able to discover thanks to this approach to the study of poverty. I am going to finish university in a year and I have to say that I had never had a subject that showed poverty so comprehensively and offered tools for its measurement and frameworks to evaluate its progress. This is the rigor that a matter of this caliber deserves, without falling into sentimentality and making me reflect from a human and deeply analytical perspective at the same time. Another point to highlight is that it may seem silly to play video games to understand such a puzzling matter, but the truth is that it is one of the highlights of the subject, you can feel some frustration and helplessness in the face of the vulnerability that governs poverty. Basically it's a great tool to get out of preconceptions. When I have children I will make them play these games.*

## Reflection #2 (student, University of Navarra, Spain)

*When someone start talking to me about poverty there is only two things that comes to my mind; social resentment and ignorance, and I have to be honest I used to be the typical person that thought that he knows everything about poverty because I grew up in a third-world country which after this subject I learnt that I didn't know anything about poverty. Finally, I have to confess that the idea of playing a game to understand poverty does not convinced me at all; but when I started playing Ayiti and I saw how complicated are to live in any country that has a similar situation to the one presented on the game I got an idea of how close poverty is and how complicated life is for that people. Spent was a game that really make me rethink of how life is in the developed world; since I lived in a developing country, I always have the dream to live in the developed world. However, you do not consider how it really is, how much money you need to spend and how unfair and complicated is to growth in this scenario. With these two games I understood how complicated poverty is in any circumstance and also, I grasp of the problems that every system has in order to end poverty.*

## Reflection #3 (student, University of Navarra, Spain)

*I personally think that before enrolling in the course we had little knowledge of poverty. Moreover, I believe that we developed a transformational understanding of different aspects of poverty and sustainability during the course. This course treated poverty and facts around it very differently than just facts, indicators or indexes. Throughout the course the Professor made us go beyond them. I could define the course as a journey through poverty and sustainability.*

*Finally, we got to the part that may seem less important but that I think that also made an impact. Playing two computer games: Spent and Ayiti. With the experience of this game the teacher made us put ourselves in someone else's shoes. One was about trying to survive a month in the US with 1000$ and the other is trying to raise a poor family in Haiti. These two experiences made us realize how difficult it is to achieve what we consider a normal life. I consider myself as really privileged in comparison to other places of the world and I think that the journey taken through the course made us realize this and also that we have a say in all of the matter. I believe that a youth aware of what poverty is and how hard it is to get out of it will try to help to eradicate it. We should have this aim and this course helps to become closer to it.*

## Reflection #4 (student, University of Navarra, Spain)

*The exercise that stood out the most to me was the game called Ayti. I found that it was the most effective in keeping you interested, all the while demonstrating the point that this "game" is in fact a reality for many people. The frustration and incompetence I felt made me realize the "solution" to poverty wasn't as simple as I thought.*

## Reflection #5 (student, University of Navarra, Spain)

*During my University years, I have taken two courses related to poverty. I enjoyed the first one a lot; it was … a very mathematical approach to poverty. I enjoyed the course … but it … gives a very partial view on the concept of poverty, which further includes social, political and even environmental aspects. Prof. Rodríguez's class helped me complement this view. She not only went through some of the indicators that I saw in my previous course, but also helped us discuss this problem in a much more humane way.*

*Throughout the course, we discussed poverty distribution in the world, utterly relevant facts (not only general but also rather specific) of the problem of poverty, and had some very dynamic group projects and games that helped us to (1) understand that we know much less about poverty than we think, and (2) put ourselves in the shoes of the population living under poverty. The latter is the one I enjoyed the most: we played two individual games in which we had to make extremely harsh life-or-death decisions which people living in conditions of poverty make every single day (also in developed countries).*

*I thoroughly enjoyed the course and recommend the journey because of its dynamism and profound learning experience.*

## Reflection #6 (student, University of Navarra, Spain)

*In our Global Political Economy class, we learned about many important topics. However, the most impactful could have been learning about poverty in our development section. We were asked to play two different games, Ayiti and Spent. For me, these were truly impactful. I was put in scenarios that maybe I knew existed but had never really thought so deeply within them. There were cases where I realized how much I would take for granted, for example from things like health and education, to even being in the position of making an immoral decision in order to have income for the health of my children, or to pay for an unseen expense. Then, to top it off we wrote some reflections on this, while we were in quarantine, where we could put into words what we actually learned. I understand that due to the situation we had to do this as a written assignment, which is great. However, what I would add, is that if possible, have students write their reflections and in small groups during class, share their work and create a discussion on the matter.*

## Reflection #7 (student, University of Navarra, Spain)

*Through this learning process in the class of Global Political Economy I have learnt the complex concept of poverty from a much wider point of view. The first activity that we took part of was defining poverty and inequality in groups formed in class. This activity gave me the idea of the complexity of trying to define the concept. While doing the activity we were discussing several factors which helped us to comprehend the difficulty of the topic. As an example, maybe one of the members of the group would say that being poor means not having "enough" to live with "dignity," but then the discussion would be: How do you actually define "enough" and how do you define "dignity"? Rather than the actual answers that we gave (which were somehow on the right direction), this gave us the insight of the complexity of defining these terms.*

## Reflection #8 (student, University of Navarra, Spain)

*These games made me feel anxious and in despair. I used to believe that getting out of extreme poverty was something easier than it really seemed after playing the games. It was just a game, but I keep feeling the pressure of seeing how my family members were getting sicker and sicker and I couldn't do anything. The despair of not being able to help your kids to get a better life and opportunities through education because I needed them to work so we can eat. The game was stressful and harsh. I feel it transformed my way of seeing the world.*

## Reflection #9 (student, University of Navarra, Spain)

*Ayiti made me realize how hard it is for a family to have a balanced life between economic stability, health, education, happiness … But I need to admit that, unconsciously, I thought this could only be a problem in undeveloped countries such as the Haiti and not in Spain (for example). Basically, in the second game (Spent), when all the bills and problems started to happen, my first reaction was: Why are all these problems happening to me in such a short period of time?????? In fact, this is what happens daily to more people than what we think. It made me realize there are many things that we can't control when they happen and that can show up in the worst moments, forcing us to make very hard decisions.*

## Reflection #10 (student, University of Navarra, Spain)

*I always thought as poverty as a difficult thing for people, that they were having struggles in life, and so on. Nevertheless, this was a view apart of reality. To show what I'm trying to say, I'll use and example, I knew that escaping poverty wasn't an easy job, but when I played the "Ayiti Game," I realized that it is almost impossible, that is way harder than what I thought.*

*Individually, I've grown a lot as a person, knowing that poverty is a "real deal," that there are people who are actually starving until death. I think that for challenging topic, we need challenging ways of learning; sometimes, the conventional path is not enough whenever someone is trying to make an impact in society. This new ways of teaching are adapting to a society that is leaving behind the magistral lesson, and all the "old school" teaching. The education sector needs ways to achieve an impact in their students.*

### *Reflection #11 (student, University of Navarra, Spain)*

*I personally love the way we covered the section. I like that we have activities, see videos and comment articles in class because they are more striking than just telling us so then I can remember facts and all more easily. I very much liked the videos of Hans Rosling which I then showed to my family and we had a long talk about them. I also liked the 13 questions about poverty, because I noticed I did not know as much, and the game where we had to live with 1000$ for the month because it made it very real for me, what relative poverty means.*

## Conclusions

The new way of teaching the topic on poverty and development proved fruitful, and informal feedback suggests the students were more motivated and engaged. Compared to previous years, students' assessments on this chapter were more positive and it climbed up positions in the questions of "which part of the course you liked better?" and "where do you feel you learnt the most?"

The combination of activities (lectures, videos, readings, student presentations, quizzes, games and student reflections) provided a diverse learning experience that kept the interest better than traditional lectures alone. In particular, the experience of including online games in the study and discussion of poverty and development has been very positive. Students report enjoying the games and being more motivated to study the material. More importantly, the games have facilitated a first-person thought process about poverty that would otherwise be difficult to incorporate in more traditional teaching styles. It would be naïve to assume that a mind shift occurred in all the students, but the reflections suggest that it did take place in some. Figure 14.2 groups the personal reflections in five categories, based on the general tone of the text, as follows:

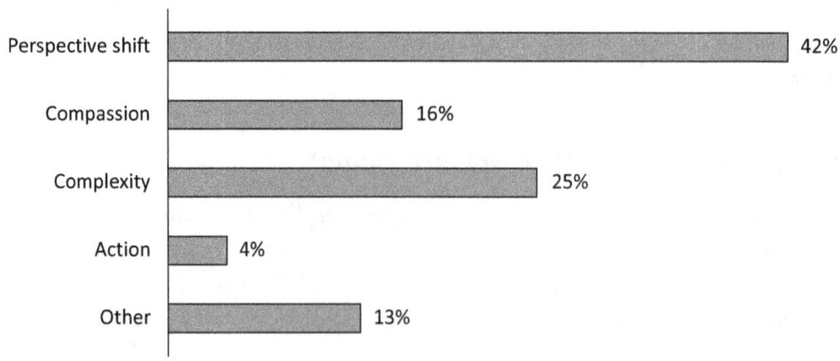


**FIGURE 14.2** Student reflections, by reaction type

(1) *Perspective shift*, where the student describes a change in how they look at poverty or increased self-awareness. Examples include "*It made me be in their shoes, that is something I don't commonly do.*"
(2) *Compassion*, where the student describes getting a better feeling of what life is like for someone living in poverty). Examples of such responses include "*theoretically, we all know what poverty implies, but it was incredibly different and more insightful to actually be the one running out of money despite doing every single thing possible. I was really shocked as, being honest, I always thought that through hard work, almost anyone could get out of poverty. This has radically changed my mind though.*"
(3) *Complexity*, where the student describes a new understanding of the *multi-faceted reality* of poverty, or the realization that different seemingly opposite perspectives can actually co-exist (both-and thinking). Examples of such responses include "*By playing the game I have realized that getting out of poverty even if you want to is more difficult than it seems, all strategies have their problems and all their benefits.*"
(4) *Action*. Here are comments where the student mentions being moved to *action*.[5] Examples of such responses include "*I want to be involved and help.*"
(5) Other-uncategorized.

Based on the experience, some precautions may be useful if incorporating these (or other GBL) strategies in a discussion of poverty in an undergraduate setting:

- The games need to be presented clearly, so students know what is expected of them going into the activity. Stating the learning goals is of particular importance, as already noted in Moncada and Moncada (2014).
- The games need to have a clear connection with the rest of the material, so they are not played and quickly forgotten. It is not enough to include them in the relevant topic; a particular effort to connect them with concrete concepts, measures and policies seems to increase their value. A synchronous session, where discussion among the students and with the instructor is facilitated to link back to the more theoretical aspects and initiate a discussion to foster reflection, proved useful.
- Depending on the audience and the tool used, the name "game" may be better substituted by "simulation." Otherwise some students may consider the material less important or not "serious." As Whitton explains "Associations with play as an activity that is childish, frivolous or inauthentic may limit the motivation for learners – particularly adult learners – to engage with playful learning" (Whitton, 2018: 9). We indeed found this in one of the long reflective essays. As one student wrote:

> *there is a risk to be considered and balanced: while these exercises were lighter and more dynamic, they also gave the impression that the subject of poverty was less serious, less important than others, and we as students attribute less importance to it.*

Given its low commitment in terms of time, including first-person games in the theoretical discussion of poverty and development seem to be an effective way to motivate deeper thinking. They also provide an easy-to-implement strategy to approximate experiential learning for a topic where students from the developed world may have limited exposure through their lives. If the activity facilitates a mindset shift even in some of the students, its cost-benefit potential is high and suggests further consideration.

## Notes

1 "Deep learning" is a concept often used to refer to student outcomes focused on the acquisition of complex, interrelated learning and high-level skills (such as critical thinking, the ability to work effectively in teams, solve complex problems and being a self-directed learner). Interested readers may get a better understanding of the concept in the context of teaching for sustainability in Hermes and Rimanoczy (2018).

2 Honest feedback was essential to gauge the impact of the activities, so students were guaranteed anonymity. For this reason, all reflections (marked in italics) are nameless.

3 Student insights are presented verbatim, except when doing so made the text difficult to comprehend. Because most of the students have English as a second language, grammatical errors or unusual wording may be present. However, this seems preferable to the possibility of altering meaning while editing their texts.

4 One of the editors of the book, Isabel Rimanoczy, provided an interesting insight. Because this type of exercises can sometimes touch on the feelings of participants and even make them uncomfortable, it may be beneficial to anticipate the possibility so students don't end up depressed, defensive or powerless. This may also be a good place to offer a possibility to channel those feelings into action. Dr Rimanoczy suggested a potential opening explanation that may be useful to others interested in using the games: "This game will put you into the shoes of a person/family living in poverty, and it may bring up strong emotions, depending on how much you take the game seriously. This is not meant to make you feel bad, but yes to feel things from other people's shoes. If you get anxious or stressed, remember this is a game. And that it also may be true. Perhaps you can convert your feelings into actions: what is one thing you could do? There may be small things that indirectly help. Not to the fictional people of the game, but to real people dealing with poverty."

5 It is, of course, impossible to know whether this impulse to action did translate into actual changes, but it is still interesting to consider that a seed has been planted. An interesting extension may be to link up the last session of the topic with an introduction of student volunteer groups or other concrete forms of action.

## References

Albuquerque, R.D., Fialho, F.A.P., Pereira, A.T.C., & Gonçalves, M.M. (2009). Digital game based learning with free online games: Analysing two games against poverty. In International Conference on Interactive Computer Aided Blended Learning, Florianópolis, Brazil.

Allcoat, D., & von Mühlenen, A. (2018). Learning in virtual reality: Effects on performance, emotion and engagement. *Research in Learning Technology*, *26*, 1–13. https://doi.org/10.25304/rlt.v26.2140

Bonde, M., Makransky, G., Wandall, J., Larsen, M., Morsing, M., Jarmer, H., & Sommer, M. (2014). Improving biotech education through gamified laboratory simulations. *Nature Biotechnology*, *32*, 694–697. https://doi.org/10.1038/nbt.2955

Crocco, F., Offenholley, K., & Hernandez, C. (2016). A proof-of-concept study of game-based learning in higher education. *Simulation & Gaming*, *47*(4), 403–422. https://doi.org/10.1177/1046878116632484

Day-Black, C., Merrill, E., Konzelman, L., Williams, T., & Hart, N. (2015). Gamification: An innovative teaching-learning strategy for the digital nursing students in a community health nursing course. *The ABNF Journal: Official Journal of the Association of Black Nursing Faculty in Higher Education*, *26*(4), 90–94.

Farrell, C. (2005). Perceived effectiveness of simulations in international business Pedagogy. *Journal of Teaching in International Business*, *16*(3), 71–88. doi: 10.1300/J066v16n03_05

Galanek, J., Gierdowski, D., & Brooks, C. (2018). *Ecar Study of Undergraduate Students and Information Technology*. Educause Center for analysis and Research. https://library.educause.edu/~/media/files/library/2018/10/studentitstudy2018.pdf?la=en

Hermes, J., & Rimanoczy, I. (2018). Deep learning for a sustainability mindset. *The International Journal of Management Education, 16*(3), 460–467. https://doi.org/10.1016/j.ijme.2018.08.001

Lovelace, K., Eggers, F., & Dyck, L. (2016). I do and I understand: Assessing the utility of web-based management simulations to develop critical thinking skills. *Academy of Management Learning & Education, 15*(1), 100–121. https://doi.org/10.5465/amle.2013.0203

Moncada, S., & Moncada, T. (2014). Gamification of learning in accounting education. *Journal of Higher Education Theory and Practice, 14*(3), 9–19.

Rimanoczy, I. (2021). *The Sustainability Mindset Principles: A Guide to Develop a Mindset for a Better World*. New York: Routledge.

Rogmans, T., & Abaza, W. (2019). The impact of international business strategy simulation games on student engagement. *Simulation & Gaming, 50*(3), 393–407. https://doi.org/10.1177/1046878119848138

Subhash, S. & Cudney, E. (2018). Gamified learning in higher education: A systematic review of the literature. *Computers in Human Behavior, 87*, 192–206. https://doi.org/10.1016/j.chb.2018.05.028

Whitton, N. (2009). *Learning with Digital Games: A Practical Guide to Engaging Students in Higher Education*. London: Routledge.

Whitton, N. (2018) Playful learning: Tools, techniques, and tactics. *Research in Learning Technology, 26*, 1–12. https://doi.org/10.25304/rlt.v26.2035

# 15

# DEVELOPING AGENCY

## Nurturing the doing dimension through the social impact scholars program

*Brian Hanssen and Manthan Pakhawala*

### Introduction

Over the past decade I've worked with hundreds of NYU first-year business undergrads in a course called *Business and Society*. Students dive deep into the SDGs, ponder stakeholder theory, consider the implications business decisions have on society, think critically about a corporation's positive and negative externalities, and call on companies to address their shortcomings. Interactive plenary sessions expose students to business leaders, entrepreneurs, academics and activists – all within the context of the SDGs. Small-group discussion sections encourage students to share their perspectives and become better writers by thinking deeper about business concepts. Above all else, we teach *thinking* about business, sustainability, ethics and social impact. And, in many regards, we are ahead of the curve integrating the sustainability mindset into our core business instruction.

Yet, too many still leave the course believing they have to make a choice: get a "real" job, pay off student debt, fulfill parental expectations – the list goes on – *or* dedicate themselves to bettering the world. The SDGs, once part of daily discussion, sometimes become a distant memory replaced by debt ratios and recruitment events. Of course, we are making progress on this front every year, but for a long time now, I've been taking it personally. Have I not done my job creating an engaging space for them to think bigger about the world of business? Do I impose too many of my own values on students that simply don't stick? Am I not persuasive enough? And worst of all: do the students just feign interest in these topics to get a good grade?

Moreover, the entire time I've been teaching *Business and Society*, I considered my instruction style to be "experiential" when it turns out I've been missing numerous components. Eyler (2009) describes experiential education as helping students to bridge classroom study and life in the world, transforming inert knowledge into knowledge in use. Smith (2002) suggests experiential learning moves the learner from dependency to self-directedness without the teacher taking full responsibility for determining what is it to be learned, when it is to be learned, how it is to be learned, and if it is to be learned. And, Duhon-Hanes (1996) argues that controlling one's *own* learning experience is crucial to the educational, intellectual and personal development of students. This research opened my eyes to my precisely planned,

DOI: 10.4324/9781003229735-17

to-the-minute timed activities aimed at keeping student *attention*, encouraging participation, and meeting *my* distinct learning objectives. While these are not bad things, I confused *engaging* with *experiential.*

In order to fully immerse myself in experiential learning – in part because I didn't want to make any waves with Administration and also because I didn't want to fail in front of my colleagues – I decided to start an entirely separate not-for-profit organization, operated during the summer months, called the *Social Impact Scholars*. The goal: group student participants according to interest, provide them with resources, partner them with a social entrepreneur (who addresses at least one SDG) to consult on a problem and transfer *agency* to them so they have control over the outcomes. I asked myself a lot of questions: How can I empower students to engage with the SDG's as an *agent* or *partner*? How can I complement the social impact work we are doing in courses like *Business and Society*? How can I better prepare students for the business world? How can I use my background and expertise to create a supportive environment without dictating the process?

Before I knew it, I was the Executive Director of a *12-week summer internship* for first-year *college students* mostly from NYU and a few other top business schools in the US. By year two, I accepted 70 student participants, or *Scholars*, interested in addressing real-world problems with impact-driven businesses focusing on at least one SDG. Below are a few of the most important program components:

- *Scholars* consult directly with a mission driven company or nonprofit in real time *and make all final decisions.*
- *Scholars* (college freshmen) work in teams of four with an upperclassman guide (who participated in the program before) *whose job is to ask questions – not give answers* as well as a mentor (typically an alumnus looking to give back) *whose job is to ask questions – not give answers.*
- *Scholar* teams work off of a high-level problem statement provided by the client. However, participants are given the freedom to develop the problem statement in ways that motivate *them* and meet the summer timeline.
- Online training modules (described in more detail later) are available in a just-in-time setting.
- *Scholars* are empowered to make decisions and recommendations on their own, and they have to engage with the clients, ask the right questions, think through the problem deeply, and come up with some sort of recommendation that actually speaks to the client. None of this is prescribed to them, and the outcome is not based on a grade or a class.
- Feedback (described in more detail later) mechanisms and reflection opportunities become essential throughout the summer projects.

## Ownership of learning

Perhaps the most important attribute of *Social Impact Scholars* centers around student ownership of problem solving and decision-making. Each team has an upperclassman lead to support logistics, provide advice for team dynamics and share stories from the past project they were a part of the previous year. Teams are also assigned a mentor in the working world (typically an NYU MBA alumnus) who provides a "reality check" for students at critical points in the project (designing a framework, data gathering and final presentation). When appropriate, mentors can give advice, but I regularly remind them to ensure the student teams are running every aspect of the project and coming to decisions and conclusions on their own. For example, when a *Scholar* recently asked me for advice on their project approach, I asked her to defend *why* her framework covers a wide range of exploration areas while still focusing deeply enough to be thorough. By the time she finished explaining her reasoning she had already thought of two new research areas!

Online training videos provide further guidance when students need more than just questions in response to their questions. Rimanoczy (2016) suggests providing just-in-time interventions, meaning waiting until the timing is right to introduce a concept, as opposed to sticking to a preplanned agenda. *Social Impact Scholars* modules include:

- Empathizing with the client
- The first client meeting and interview
- Frameworking a problem following the case consulting process
- Data gathering and research
- Design thinking and prototyping
- Presenting final recommendations
- Dealing with team conflict
- Facilitating meetings
- Consensus building.

These resources are short, typically four to five minutes in length so that students feel more comfortable accessing and applying the basics on their own. I find that long readings and specific case studies can be daunting for many students, who either get lost in the material (without proper guidance from the professor) or simply ignore it entirely. This strategy represents a hefty departure from some of my traditional courses, some of which require a book or more of reading a week, but it's absolutely worth it in an experiential learning environment; students are more willing to engage with the information and, over time, recognize that an "expert" will not be providing explicit direction.

## Connection to social entrepreneurs

*Scholars* are encouraged to form a connection with the client and get to know them as a person. Such conversations provide students with a model through which to ask important questions that cultivate the sustainability mindset: Who am I? What is my purpose? And, what difference am I making? (Rimanoczy, 2010). These are important questions! Rimanozcy (2016) offers that many of the business leaders she interviewed felt they would have experienced earlier personal development and growth if they were exposed to [such] challenging self-awareness provoking questions during their college years.

## Finding one's own value

Ward and Wolfe-Wendel's (2000) review of literature on service learning reveals the campus and its constituencies often benefit more than the community (in this case, the client). This has also been a concern and sticking point with *Social Impact Scholars*; I, too, have received criticism that some of our projects fell short of a "real" consulting project. Most of the *Scholars* population, first-year college students, lacks the necessary training to run sophisticated financial models or design a detailed marketing strategy, for example. And yet, this is one of the most important opportunities we have to instill *agency*. By constantly asking participants, "What value do YOU bring to this project?" I re-calibrate the students' thinking around what *they* are experts in. Typically, this has a lasting effect on project quality: that same group of students who couldn't put together a strong marketing plan *can* interview 20 marketing professionals, break down and analyze the conversations, and provide *solid insights* clients can use. And, in the process, *Scholars* gain confidence and clarity around how to frame their value and think through any challenges future bosses will throw at them.

To accomplish this important step, it is important to frame the conversation properly with clients. These types of consulting projects work best when students are collecting *primary source information*: interviews, surveys, focus groups, product testing, etc. First-year students are quick to rely on speedy Google searches with little or no context – typically not helpful for clients. But when you push the student to sit in the driver's seat (Freshmen hate cold-calling!), she or he exercises an important skill most of us never even touch upon in the classroom.

As importantly, it is important to position *Social Impact Scholars* projects as unique opportunities for clients to better understand the upcoming generation of workers, their communication style, expectations from management and general working preferences. Because *Social Impact Scholars* has such a robust feedback mechanism (described in more detail below), clients get a much more realistic picture than they would from young employees working for them directly.

## Problem solving in "sets"

Perrin (2014) notes that ownership increases when students are responsible for choosing their project or strategy for change, identifying and solving *problems* rather than relying on a teacher or supervisor. While *Social Impact Scholars* clients do propose a high-level problem statement for the students to tackle, it is the *Scholars* themselves who are required to define and scope the problem and project themselves. Two examples of high-level problem statements and student reframes are below.

Table 15.1

Again, by asking students to consider their *value* and realistically scope the project in ways that leverage this value, framing problem solving becomes second nature.

When designing the *Scholars* group teams (typically four or five students), I was also inspired by the concept of action learning "sets," described as free-flowing discussions based around a presenter's issue – yet leaving the responsibility firmly with the presenter (McGill & Beaty, 2013). Each team member has a chance to present their challenges to their (first-year and upperclassmen) peers, with their experienced mentor, to me, and in some cases the client. The common characteristic is that instead of organizing a session around content that has to be taught, it is organized around problems that have to be solved (Rimanoczy, 2016). Since feedback is such a big part of the program, these learning teams lead to tremendous self-awareness for both the *Scholars* and the clients. And, since *Scholars* both identify the problem and are held accountable for addressing it, they become empowered to follow a similar process in situations they face after the program is finished.

**TABLE 15.1** High-level problem statements and student reframe statements

| *Client problem statement* | *Student reframe statement* |
|---|---|
| "How do I grow my composting business in New York City?" | "Where in Brooklyn would be the ideal spot to host compost drop-off locations?" |
| | "What are the most effective marketing techniques to attract people to these drop-off locations?" |
| "Assess the possible means of establishing a steady stream of revenue for our nonprofit, including utilizing the donor outreach solution ('sustaining member base')." | "Which top CRM platforms are optimal for non-profits and donor outreach (using SWOT analysis and other business frameworks)?" |

## Feedback and reflection

A critical factor for achieving powerful experiential learning outcomes is the inclusion of opportunities for feedback and reflection (Eyler, 2009). *Scholars* are required to give anonymous feedback on a weekly basis to their (1) peer team members, (2) Project Lead and (3) client. They also participate in two to three programmatic surveys and/ or interviews throughout the summer (reviewed by an internal project team dedicated to improving our *own* program). Boud (1995) describes this feedback as extremely important because it allows students to strengthen their ability to self-assess through assessing others. And, by being held accountable by their community partners (clients), students produce work that matters (Perrin, 2014). Perhaps more importantly, this feedback provides an opportunity for students to reflect and think about their own purpose, their strengths, their weaknesses and how they will contribute to society in the future. Feedback also allows students to work on their communication skills, as they must effectively articulate what made sense and what didn't (Liu & Carless, 2006).

Two of my favorite survey responses give me hope that this feedback process is working:

Student: "*I feel like I am being exposed to the right resources and people in order to develop as person.*"

Client: "*This was exactly what I needed at my phase in the startup process. Given that I'm a one-man team for now, just having a team of students working made me feel 100% more productive with my time. I felt very heard the entire time and everything I mentioned was addressed and executed on.*"

## Outcomes

Ultimately, *Social Impact Scholars* attempts to embody what Perrin (2014) recommends for student learning: the capacity to perceive and address ill-structured problems, tolerate ambiguity, make warranted judgments and act while continuously seeking and refining further information … through questioning, commenting and challenging one another in an informal dialogue. As importantly, students develop their "being orientation" (Rimanozcy, 2017) while exploring and understanding their roles as business leaders, consumers and members of the planet Earth.

Ultimately, *Social Impact Scholars* pushes students to go deeper in their knowing/ thinking, doing, *and* being. Rimanoczy (2010) suggests that these deep questions, such as the purpose of one's own life, the meaning of work, one's own role, personal values and assumptions and other spiritual questions should be a core contribution of management education. The *Scholars* program asks these questions at the time when students need them most!

## Applicability in college classrooms

Of course, setting up a nonprofit organization and developing the infrastructure, training and networks (and running it throughout the summer) may be too much of a task for most educators. However, running this program outside of the classroom has inspired me to rethink *all* of the teaching I do *inside* the classroom, and I hope to have provided a model that can actually be replicated within the university setting. For example:

- Connect students together in small, semi-autonomous working groups. If possible, let the students group by interest.
- Instruct clients (social entrepreneurs) to provide high-level problem statements that require *primary source data collection* as a means of solving them. Clients should be open to student participants being able to manipulate the problem.
- Pose questions and ask students to tackle them instead of dictating a prompt.
- Ensure students have agency in understanding their value, determining their outcomes *and* defining success metrics.
- Position support structures as advice and question sessions.
- Post online resources (i.e. flip your classrooms?) with just-in-time information instead of dictating a totally laid out set of expectations, guidelines and weekly readings.
- Develop relationships with business leaders and mentors in ways that express a *partnership* instead of a one-way teacher/ student relationship.

Students from both the *Social Impact Scholars* and my university courses have expressed incredibly positive feedback with this approach. I asked one such student, Manthan Pakhawala, to share his first-hand perspective of how *Social Impact Scholars* (and the similar approach I have started to take in the *Business and Society* course) shaped his thinking.

## Student reflection: The dichotomy of intrinsic motivation and external action

Climate change represents the single largest challenge that humanity might be facing, a challenge that would take the combined cooperation and effort of all the afflicted stakeholders. However, I have also noticed an incredible disconnect between the awareness of the challenge and the action that would be needed to work towards solving this. A University of Florida Institute of Food and Agricultural Sciences (2013) study highlights how college students are worried about climate change-related hazards, even if they're not worried about climate change. College students still perceive the threats of climate change as theoretical, thus representing a gap in their knowledge and their actions.

As a business student with a secondary background in psychology, I understand the theoretical difference between internal and external motivation. Despite knowing the value of the SDGs, the constant need to start addressing the issues from the grassroots level, and the level of impact I could make as a business student, it is really tough choosing the right actions that could match that knowledge. I know sustainability is important, but so is securing a financial future as a first-generation immigrant. I know that the threat of climate change is much more permanent, but the non-visibility of the challenges aids in the illusion of it not being there. I have also been able to compartmentalize the two versions of myself in my brain – an idealistic one that cares about the SDGs and a present one that goes with the flow of going through the motions of grades and jobs.

It was not until my first summer interning with the *Social Impact Scholars* that I began to reconcile the two tensions of idealism and realism as well as understand the value of fostering internal motivation for long-term adherence to the SDGs. Placed as an intern on a team of five, we were assigned to consult for a cryptocurrency client who wanted to leverage blockchain technology to generate venture capital for African businesses. Throughout the summer, we struggled to make consistent meetings with the clients, show up to weekly meetings with our assigned work done and also find a convenient time for all of us to meet – something we failed to do surprisingly often. Even though we managed to present a project that met the client requirement at the end, I realized that the end product was not the best it could have been. It was especially disappointing since this was a client and a mission I really cared about. I come from a comparably developing economy and raised in a family that valued entrepreneurial drive and innovation. I also knew the value that such an organization could have added to the lives and people of the market it would enter.

The difference in my value system, where I valued the work my client was doing, and the eventual result of a deliverable that could have been better, forced me to take a look at what could be driving force behind the difference. This is when I realized that a key difference between my prior experiences, which were primarily based in and around school environment and my summer internship. I realized that I had grown accustomed to working for a grade, analyzing the key metrics and understanding the grades' exact boundaries. This helped me plan out my semester, where I knew exactly the amount of work I had to put in to get the desired results. It was a strategical algorithm that optimized my chances of getting an A. However, the system relied a lot on enforced and continuous deadlines, the peer pressure of adhering to the said deadlines and constant feedback that I would receive after every assignment submission.

On the other hand, the key part of the summer program was that it was designed to be what the interns make of it. The project's quality relied on my willingness and ability to research, synthesize, and come up with more profound and more meaningful insights. Since the program was also unpaid, there are significantly less "external" stakes and rewards. Instead, it forced the student to rely on an internal motivational system that the educational system doesn't do the best job of fostering. During the summer, I could not rely on an algorithm that maximized by reward with the least effort. Instead, I was left with the task of confronting my idealistic self and having it work without an expectation of immediate feedback or worth.

The next summer, I had the unique opportunity to not only right my mistakes, so to speak, but also effect change at a deeper level. I was given the position of Team Lead for SIS, a leadership position that puts me in charge of a team of four to five consulting interns for the duration of summer. I was tasked with guiding and managing the group – something that I was really excited about since I saw myself in their shoes. I wanted to use my introspection, findings and feelings from the first summer and immediately put it into practice. I wanted this summer to be my redemption arc.

It turns out that looking at the prospect of managing a team as an opportunity to right your previously conceived wrongs is not the best style of leadership, especially for a budding group of rising sophomores that have their own unique expectations and goals they want to achieve. Because of this mismatch, the first month of my experience as a team lead represented a significant learning experience for me.

I started to micromanage my team since I wanted fix my mistakes immediately. I wanted to do the project for them. I would sit in on their meetings and rather than allow them to brainstorm and come up with their own solutions, I would be compelled to share my version of the path that they should take. This version of leadership style also meant that rather than offering myself as a resource in times of help or distress, I was effectively doing the project for them – treating my consulting interns as appendages I needed to control to reach the goal in the way I thought was the most optimal. I would also assign specific tasks for each member that they would need to complete by our next check-in meeting.

I noticed something was amiss when I had to miss a check-in meeting due to personal reasons. When I checked in with the group a week later, I was surprised to note that the group had made no independent progress in the previous week. Instead, they waited for me to review their work and assign the next work that would need to be done. This lack of initiative stemmed from my inability to separate myself from the project. I realized that I had invariably created the very ecosystem that created the two system dichotomy of intrinsic motivation and external reward based drive. Instead of fostering their internal drive for real change, I had replaced the school structure of strict deadlines and feedback with my feedback and deadlines. The experience helped me understand the fine line that I would have to tread in order to foster the internal drive that would help the students of my generation connect the knowledge of SDG goals to actionable output in support of that knowledge.

I had to learn how to take a step back in order to minimize my micromanagement and foster the feeling of ownership. I learned how to do this by delegating roles and responsibilities but also creating boundaries by not asking for or checking for work. The weekly meetings, which previously were a space for "submitting" work for feedback, became a space where the interns could meet, discuss and deliberate on the next steps that they wanted to take in the project. I also strived to reposition myself from a moral authority that checked their quality of work to a supportive resource, available to help them achieve their goal.

I learned how easy it is to fall into the trap of micromanagement when you have a specific objective that you want to be implemented. However, the style also breeds the separation of intrinsic motivation and external behavior. My interns were no longer motivated to provide the best project that they could have and instead, they were now aware of the one metric that they had to hit in every meeting – my approval.

Fortunately, I started noticing a marked difference after my efforts to delegate and step back. The team was now showing more initiative and actively planning meetings to discuss the project without taking my availability into question. This was possible because they understood that now I wasn't a metric that they had to satisfy anymore. Another important factor was the presence of a client that they felt passionate about. The client wanted the team to explore the vertical farming industry in the United States and believed that vertical farming is the next step in the farming evolution. The sustainable and productive benefits of vertical farming aligned with the interns' internal need for more sustainable business practices. Thus, the client work served as a bridge to connect the intrinsic motivation to their external actions. The project provided the interns with an opportunity to extend the values that they believed in through their professional endeavors. The lack of specific target also helped foster a sense of internal drive for the project, something that I lacked for my first summer with SIS.

The team ended up turning an incredible project, surprising me with their depth of research. They also asked extremely relevant questions and went a step further by actively engaging the client and asking counter-questions when necessary. A key question that stood out to me was:

"What value add do you think this recommendation is going to have for your organization?"

The question meant that the team was not just meeting the project deadline but was genuinely concerned about the project beyond the scope of the summer.

The experience taught me the value of the right leadership and also stressed how the role of the team leader is to inculcate the feeling of ownership in the interns.

The first summer internship with SIS encouraged me to understand how every piece of information and technology exists in relation with one another. I think it is a human fallacy to try to categorize and, thus, box the disparate concepts into their buckets. Consequently, despite being aware of the concepts of "blockchain," "venture capital" and the "African continent," they existed in separate categories inside my mind, invariably oversimplifying the concepts. I had never imagined that a concept such as blockchain, which occupied a highly buzzwordy and technical place in my brain, could be used in relation to venture capital, another word that I had only connected to entrepreneurship. The client's ability to not only merge the two worlds but also use the unique advantages of the blockchain technology to leverage successful venture capital shattered my disparate world view of the two concepts.

Each new generation represents a new hope for our civilization to move forward. Although it might not seem like it at first, humans have a natural instinct to preserve their environment for the future generations. But, it might be tough to engage with the problem without fixing the instinctual biases that make it tough to engage with these issues. The university system presents a challenge of the two selves. The "informed self" cares about sustainability goals and global issues, but is resigned to an internal motivation that does not necessarily take action. The extrinsically motivated self is taught to seek continuous rewards for their behaviors through grades and high paying jobs. The SIS summer internship taught me the meaningful difference between my own two selves, and how important it is to merge and maintain the intrinsic motivation with the self that effects change in the world around us.

## Conclusion

Manthan's story so nicely articulates the struggles and challenges my students go through and it exposes the unintended effects of a "teacher-centric" classroom. Ironically, by stepping back and thinking of my students as coworkers, my classroom as a nonprofit organization, and my assignments as a client project, I learned the most about what being a good teacher *in* the university classroom can be. Perhaps that's how we can best prepare our students to tackle the sustainability challenges that lie ahead in the business context and develop the "being" part of the sustainability mindset so critical to defining one's own sense of purpose.

## References

Boud, D. 1995. *Enhancing Learning Through Self Assessment*. London: Kogan Page.

Duhon-Haynes, G.M. (1996). Student empowerment: Definition, implications, and strategies for implementation. Paper presented at the Third World Symposium, Grambling, LA (ERIC Document Reproduction Service No. ED 396613).

Eyler, J. (2009). The power of experiential education. *Liberal Education, 95*(4), 24–31.

Liu, N.F., & Carless, D. (2006). Peer feedback: The learning element of peer assessment. *Teaching in Higher Education, 11*(3), 279–290.

McGill, I., & Beaty, L. (2013). *Action Learning: A Practitioner's Guide*. London: Routledge.

Perrin, J. (2014). Features of engaging and empowering experiential learning programs for college students. *Journal of University Teaching and Learning Practice, 11*(2), 2.

Rimanoczy, I.B. (2010). *Business Leaders Committing to and Fostering Sustainability Initiatives*. New York: Teachers College, Columbia University.

Rimanoczy, I. (2014). A matter of being: Developing sustainability-minded leaders. *Journal of Management for Global Sustainability, 2*(1), 95–122.

Rimanoczy, I. (2016). *Stop Teaching: Principles and Practices for Responsible Management Education*. New York: Business Expert Press.

Rimanoczy, I. (2017). *Big Bang Being: Developing the Sustainability Mindset for a New Planetary Imprint* (2nd ed.). Sheffield, UK: Greenleaf Publishing.

Smith, M.K. (2002). Malcolm Knowles, informal adult education, self-direction and andragogy. Retrieved July 16, 2020 from https://infed.org/mobi/malcolm-knowles-informal-adult-education-self-direction-and-andragogy/.

University of Florida Institute of Food and Agricultural Sciences. (2013, October 3). College students worried about climate change hazards. *ScienceDaily*. Retrieved July 16, 2020 from www.sciencedaily.com/releases/2013/10/131003105700.htm.

Ward, K., & Wolf-Wendel, L. (2000). Community-centered service learning: Moving from doing for to doing with. *American Behavioral Scientist. 43*(5), 767–780.

# 16

# MILLENNIALS AS CHANGE MAKERS SUPPORTING SDGs

*Amelia Naim Indrajaya and Isabel Rimanoczy*

## Introduction

A major concern of educators is the challenge of engaging the interest of students, particularly those in college. This is important notably when many institutions are bringing contents related to sustainability into their programs, with the aim of preparing the new generation to address the increasingly urgent problems of our planet (Kassel, Rimanoczy & Mitchell, 2018). In 2015, the United Nations Global Compact announced the Sustainable Development Goals, a list of 167 targets grouped into 17 categories, comprising a deliberated list of all that has to be solved for a world that works for all, in the words of then Secretary-General Ban Ki Moon. This chapter presents an initiative designed and implemented at Sekolah Tinggi Manajemen Ipmi, known as Ipmi International Business School, in Jakarta, Indonesia, with the aim of introducing students to the SDGs, and engaging them in actions to create ripple effects of change.

In the spring of 2020, Amelia, one of the authors, designed an online workshop to serve as a learning experience for her first-year students in the Business School. As a member of the Principles of Responsible Management Education Working Group on the Sustainability Mindset, she consulted with Isabel, the Working Group Convener, to design a learning event for her 33 students. In previous years, both authors had run classroom activities with each of us facilitating from two different continents. On those occasions, Amelia was in the Jakarta classroom with the students, while Isabel participated via Zoom from her home in Florida, a time zone 12 hours behind Indonesia. The experience of having a participative learning class across the globe was much appreciated by the students (Indrajaya, 2018).

The COVID-19 pandemic intervened and in March 2020 we had to start online classes for everyone. Amelia decided to introduce the students to the SDGs in an online workshop, where the facilitators were situated in their different continents, but this time with students joining from the comfort of their home.

DOI: 10.4324/9781003229735-18

The class is Business Ethics and we defined learning goals:

- To make the students familiar with the SDGs
- To inspire them to become agents of change in pursuit of SDG goals
- To support them to engage in change actions.

We initially planned the class for the 33 students of the Business Ethics course. As educators, we see the students as future business leaders, and we hope that they will develop businesses in ways that support the Global Goals.

Even though the session had to be virtual, we wanted to ensure that student engagement and participation were high. We determined that this online class would not be a traditional webinar, with the teachers making a presentation while the students listen and just take notes.

Working globally presents hurdles. The timing and length of the class had to accommodate the Indonesia/ USA time zone differences plus religious observances. The class was planned in the midst of Ramadhan, where the predominantly Moslem students of Ipmi would be fasting during the day. For this reason we decided to organize a 90-minute webinar in the evening after the breaking of the fast, Indonesian time, and early morning in Florida.

## About the institution

Ipmi International Business School is a boutique business school, where classes are fully conducted in English. The school was developed back in 1984 as an initiative of the late Bustanil Arifin, former minister of Cooperatives of Indonesia, to address the fact that a growing number of students had to go abroad to earn an international business degree. He decided that it should be possible to provide an international business education in Indonesia which could attract international students to travel to, and study in Indonesia. As a result today in the Business Ethics class there are students with a nationality from Japan and Colombia.

The Center for Sustainability Mindset and Social Responsibility in Ipmi had made the Global Goals (SDGs) an important component to be embedded in value-based subjects such as Business Ethics. We realized that other schools would also benefit from more exposure to the SDGs. Therefore, Amelia decided to expand the learning experience to a broader audience and posted invitations to other young students interested in becoming change makers for a better world, setting a cap at 100 participants because of technical and design constraints.

We were pleasantly surprised that 141 students signed up to participate, although we had to limit it to 100 participants due to the technical Zoom limitation.

## Promotion

To make the event available to a broader audience, we designed a flyer announcing the class, and opening it up to students from high schools and undergraduate degree college students from Ipmi International Business School and other universities. The flyer was designed to awaken interest and motivate the students to participate in becoming real agents of change. The invitation to the event was shared through social media ten days before the event,

**FIGURE 16.1** The brochure of the Change Maker webinar. Source: Amelia Naim Indrajaya personal document

and a Google form for registration was made available through a link in the brochure (see Figure 16.1).

The title of the event "Millennials as Change Makers," was deliberately selected to attract a young audience, mostly college students. The posting of the program through social media made the invitation go viral and reach a global audience. Some of the participants signed up from overseas countries, including countries in Africa, and India. The physical distance was no longer a problem.

## Materials

The Google form also had two worksheets to be used by the participants during the webinar (see Table 16.1 as an example of a completed worksheet). The first worksheet invited students to list some initiatives within their area of control, meaning things they could do to support the SDGs in their daily behavior, something they could fully control. The second worksheet

**TABLE 16.1** Sample of filled up worksheets of area of control and area of influence

**AREA OF FULL CONTROL:**
I decide to do something, and I do it.
Others will benefit from it
I benefit because I reach my goal
Example: All the areas of my daily behaviors that depend on me. What I use, how I use it, what I shop, etc.
I will
★ *I will try to reducing plastics use and aware to the environment*
★*I will implement zero waste in my daily life*
★ *I will educating myself to be more empowered*
★*I will use the electricity wisely*
★*I will eat healthier foods to be more aware of my health*
★*I will use water as I need and be more responsible*
★*I will bring my own goodie bag to reduce plastic use*
★*I will not shopping too much stuff that I do not really need and be more responsible*
★*I will obey to the law and not discriminate people by their gender, ethnicity, skin color, and backgrounds*

**AREA OF INFLUENCE:**
I identify something that should be done, that we could do, but I cannot do it alone. So I talk to others, and we jointly plan how to do it. Or I influence others to see the need. I educate, communicate, ask questions or reach out.
Examples:
Create the Sustainability Oath
Create a student group at my school or join if already existing and tackle some project together
I will
★*I will maintain my online platform to raising awareness about gender issues*
★*I will make a podcast about general issues to educate people about certain issues*
★*I will make a campaign about global warming related to gender*
★*I will open charity to help children on the street survive during this Covid-19*
★*I will always highlight certain issues that booming on my own social media*
★*I will work together with my international partner's organization to held global movement*
★*I will work together in my grassroots community to reach wider audience*
★*I will create youth friendly toolkit to held a campaign.*

Source: Isabel's template

invited them to list things they could do that were in their area of influence, meaning things that could be accomplished with outside help, but which the students could not do alone. Hence, the students were invited to create or join in an initiative with others. As they registered to participate in the webinar they were asked to print and have handy the two worksheets to write on during the class.

## Logistics

Zoom was selected as the platform for the session. Participants were asked to have a working camera and a good Internet connection. In contrast to other online courses where the video is off, we wanted everyone to be able to see each other, to increase the personal connection and make it a richer experience for all.

It is known that the Indonesian culture has a relaxed sense of time and that students do not always show up at the scheduled hour announced for an event. This could pose a problem since the logistics of the design required the use of breakout rooms, and the technical constraints make it difficult to add latecomers into existing rooms. For that reason, we announced that the class would start at 7 pm, and the virtual doors would close at 7:25 pm, for a punctual start at 7:30 pm. In addition we set a cap of 100 participants. Our first surprise was that by 7:15 pm we had already reached the limit of 100 participants, and the class actually started early instead of after the hour announced!

## Design of the session

The session was designed as a highly interactive experience, with little lecturing, and we used images and a PowerPoint, to reinforce the oral message for students that might not be fluent in English.

With the inspiring title of *Millennials as Change Makers: Let's see what we can do!*, the session started with a brief welcome and introduction of the overseas facilitator, followed by a description of the agenda based on the conceptual model of Bernice McCarthy (1990) which describes four types of learning styles. Adapting this model the plan for the session was described responding to the questions: What is this about? Why is this important? What will we do? How will we do it? To what end? This approach created upfront engagement of the audience independent of their learning style since they had their question covered in each of the answers.

- *What is this about?*

Action on the SDGs

- *Why is this important?*

Change starts with self, but it creates ripple effects and has wide implications. Awareness of our potential is the first step towards change. We all can do something to shape the planet we want.

- *What will we do?*

Explore the challenges we have at a planetary level. Discuss SDGs and the 167 targets. Explore some targets for each of the SDGs. Reflect on our personal roles/contributions.

Identify one area in which to change a personal habit or influence others.

- *How?*

Individual reflection, work in small teams, dialogue/exchange in a large group.

- *To what end?*

Leave feeling empowered and inspired.

## Launching the activity

### *First exersice: Map of the Planet*

The first exercise was called *Map of the Planet*. The purpose of this activity was:

1) To have them reflect on the planetary challenges, grouped into themes and to tap into their knowledge about our current reality.
2) To collectively learn from each other.
3) To identify the interconnections and complexity of the planetary challenges.

We projected on the shared screen a slide containing a listing of categories of challenges or themes based on the SDGs, and asked them to take a picture or a screenshot of the table because they would need to know what theme was assigned to them after being divided into numbered break out rooms of three participants each (see Figure 16.2).

Participants had to assign one note-taker and one time-keeper, and in seven minutes come up with all the keywords related to the assigned challenge or theme (see Figure 16.3).

Using the Zoom break out rooms, we divided the audience into 33 trios. After seven minutes the groups automatically closed and the trios joined the main session. At this point, we had prepared a slide with the titles of the themes. Using the main screen, we invited the participants to call out some of the keywords they came up with, related to the challenge that was assigned to them. The facilitator wrote those words into the slide that was visible to all. Once we had enough words for all the themes, we asked them to indicate which ones were connected, and we drew arrows following their instructions (see Figure 16.4).

### *Second exercise: From doomsday to empowerment*

The picture of the multiple planetary problems and the complexity of their interconnections can easily create anxiety and feelings of being overwhelmed. To transform those emotions into action we had to first bring some sense of confidence and empowerment to the room.

**NOTE YOUR ASSIGNED THEME**

| GROUP | THEME | | GROUP | THEME |
|---|---|---|---|---|
| 1-16-31 | POVERTY | | 9 24 | EMPLOYMENT |
| 2 17 32 | HEALTH | | 1025 | URBAN LIVING |
| 3 18 33 | PRODUCTION | | 11 26 | CONSUMPTION |
| 4 19 | SOILS AND FOOD | | 12 27 | WATER |
| 5 20 | WILDLIFE | | 13 28 | CLIMATE CHANGE |
| 6 21 | ENERGY | | 14 29 | GENDER EQUITY |
| 7 22 | PEACE AND JUSTICE | | 15 30 | LIFE UNDER WATER |
| 8 23 | EDUCATION | | | |

**FIGURE 16.2** The assigned theme. Source: Isabel Rimanoczy

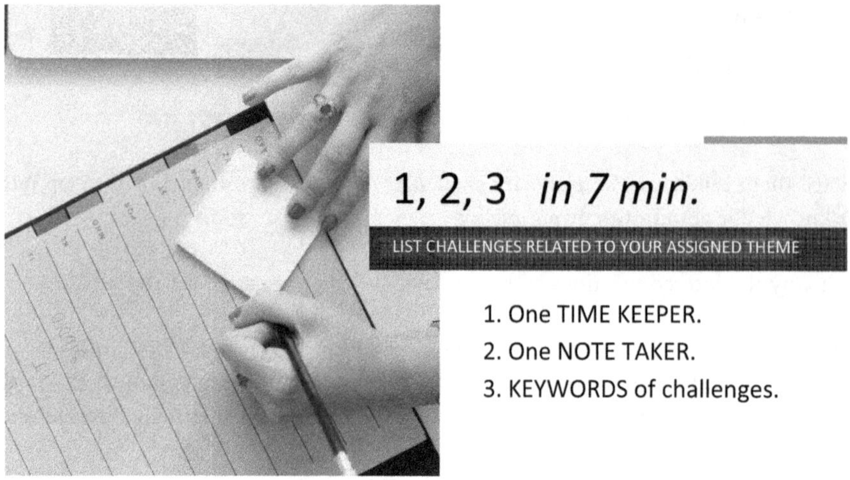


**FIGURE 16.3** Rules of the game. Source: Isabel Rimanoczy

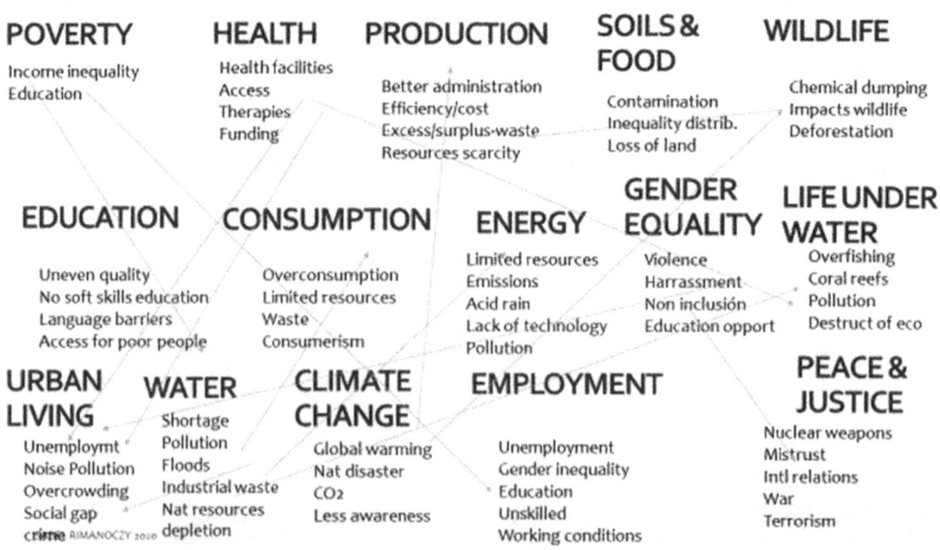


**FIGURE 16.4** The interconnected themes. Source: Amelia and Isabel's Zoom's White board exercise

We did this with an exercise where the facilitator asked members of the audience to think of a person who meant a lot to them, and who had influenced who they are today – with one condition: that person didn't know about the influence. This could be because they never mentioned it to that person; or perhaps they never met that person (the grandfather of a friend, for example); or it might be someone who had already died; or was a famous person living in another country, etc. The key here was to identify someone who played an important role in the students' life, in a positive way, and without knowing it. They had 60 seconds to reflect, in silence. Then the facilitator explained that in a similar fashion, each of us might be influencing someone without being aware of it. The things that we have control over few:

decisions that only depend on us, but our areas of influence can be very large, so large that we cannot know the limits. Our actions, words, interactions – all are small ways to influence others we may never meet.

### *Third exercise: Introducing the SDGs and some of the 167 targets*

With this sense of empowerment, the facilitator presented a slide of the 17 SDGs, and showed a selection of targets for different goals. We had previously selected the targets, seeking only those that could be relevant to the audience. Many targets are meant for government, for corporations or NGOs, and it makes it impossible for a student or regular citizen to relate to them. Some targets however can speak to individuals, for example limiting food waste, supporting local food producers, limiting energy consumption or renewable energy, protecting wildlife, natural resources, etc. The facilitator instructed participants to write down on their worksheet the possible areas in which each participant could see him or herself taking some kind of action. The purpose of this instruction was to convert the slide-show into an active participation moment since participants were required to watch, listen, notice and write down thoughts.

### *Fourth exercise: Time for action!*

The participants used the worksheet and the PowerPoint of the targets to complete their list of what actions of control and influence they wanted to take. After ten minutes, they were broken out into duos and trios to share with each other what they had come up with.

### *Fifth exercise: Sharing*

The break out groups were closed after five minutes and we invited a few to share with the whole group their commitments and initiatives. The professor asked everyone to post their commitments on their Instagram, and after a week, those that had the largest number of likes would receive a prize (see Figure 16.5 The posting with the highest number of "likes"). This strategy was aimed at:

1) using a platform for sharing ideas and feelings that the participants feel familiar and comfortable with
2) multiplying the impact of their actions by discussing it with their friends on social media
3) potentially inspiring others
4) possibly strengthening the chances of living up to their intentions because they had stated the intentions publicly
5) rewarding the "likes," thus reinforcing the likelihood of resulting in the ripple effects.

### *Sixth exercise: Feedback*

As a closing, we asked participants to evaluate the webinar by writing into the Chat box a number from one to five (one : no value, five : excellent) and indicate why. We received 100 percent of fives, not many explained the reason for their score, but those that added words made very positive remarks (see Table 16.2).

**FIGURE 16.5** Posting by webinar participant. Source: Amelia's personal document

**TABLE 16.2** Some participants' feedback from Millennials as Change Makers webinar

| *Name* | *Evaluation* | *Comment* |
|---|---|---|
| Student 1 | 5 | Because it was influencing us to realize that many problems are still needed to take action for us millennials, also it is very active for us to think innovatively |
| Student 2 | 5 | Precious material from Ms. Isabel to increase our awareness to support SDG ...big thanks to u Isabel ❤ |
| Student 3 | 5 | It inspires me to think more of small changes that can be made by us individually, and to know that the impact is big and influencing others too. |
| Student 4 | 5 | Inspired and motivated. Thank you so much for this wonderful webinar |
| Student 5 | 5 | The session is very insightful, and interactive thank you for miss isabel you are amazing, and thank you IPMI for providing us this session |
| Student 6 | 5 | The quality of the content is really great and i love how i can meet new people and discuss |
| Student 7 | 5 | I like it because it is interactive, I understand the topic and Miss Isebella is lovely :) |
| Student 8 | 5 | Because you dont only hear speaker said but i also can do anything and discuss many things with the others hopefully someday ipmi can conduct more event like this |
| Student 9 | 5 | Interactive and fruitful. I'll suggest "sutori" for interactive presentation. |
| Student 10 | 5 | I felt it was really fun, I got listen to other peoples points of views and share and at the same time game knowledge. The down side of it was few people shy |
| Student 11 | 5 | Love it and impactful |
| Student 12 | 5 | Very active and fun |
| Student 13 | 5 | Nice explanation and clear enough to understand |
| Student 14 | 5 | Because this content is very interesting! |
| Student 15 | 5 | It's interactive and interesting |
| Student 16 | 5 | Lovely discussion |
| Student 17 | 5 | Thank you, very nice explanation |

Source: Amelia Naim's personal forms

## Commitments

Over the following two weeks, the facilitator's inbox was flooded with emails attaching the individual commitments. We received 56 commitment sheets (see Table 16.3 Area of control).

The initiatives of the students in the area of full control can be classified into the following categories:

- Energy conservation
- Environmental initiatives (reduce, reuse, recycle, etc.) and
- Others (using resources wisely, and wisdom of using technology or being eco-friendly, compliance with the regulations, beneficial for others, self-development).

**TABLE 16.3** Area of control

| *Themes filled up by 63 respondents* | | |
|---|---|---|
| *Energy Consevation* | *3R (Reduce Reuse Recycle)* | *Others** |
| 38 | 39 | 46 |
| 68% | 70% | 82% |

Source: Amelia's document

**TABLE 16.4** Area of influence

| | *Main areas of Influence based on 63 respondents* | | | | |
|---|---|---|---|---|---|
| | *Family* | *Work* | *Relationship* | *Online* | *Regional* |
| Total | 12 | 19 | 36 | 19 | 43 |
| Percentage | 21% | 34% | 64% | 34% | 77% |

Source: Amelia Naim's documents

In the area of influence, we are using the leadership mandate approach (Black, 2013) by categorizing the influence into five areas of impact: Family, Work, Relationship, Online and Regional (see Table 16.4).

We have no data to permit a follow up to measure and track the commitments made by the 100 participants of the session, or the 56 that sent us their reports, beyond what was shared in this paper. However, Amelia wanted to enhance the impact potential of this session and knew much more could be achieved. So she created a follow-up step with her students.

## Taking it further

Amelia suggested to the 33 students in her Business Ethics class that they adapt the experience and create their own educational campaign targeting high school students and facilitating it themselves. The students were very excited and loved the idea. They decided to design a virtual session inspired by what they had experienced, for high school students of grade X and grade XI (17 and 18 years old). The high school selected was Al Izhar High School in Pondok Labu, south of Jakarta, because the founder of this school is also the founder of Ipmi International Business School. This created an immediate sense of trust, and the students were granted permission to invite approximately 260 teenagers to join the campaign.

They planned a 120-minute virtual session using Zoom. They maintained the goal of introducing the SDGs and invited the high school students to take action steps, which would signal their intention of becoming change agents. However, they used their own creativity to design the session in fun and new ways.

Being members of a Business Ethics class, we had a gloomy discussion about unethical conduct in the country. The students commented that unless we make Indonesia clean and free from corruption, which is the core of SDG # 16 (Peace and Justice), it would be very hard to achieve all the other SDGs. The students agreed that leaving the office of the

Commissioner of Anti-Corruption of the Republic of Indonesia (KPK) to do the work alone, would make it an impossible mission. Everyone had to join the action and work hand in hand as anti-corruption agents!

This led the students to decide that they all would work to support SDG # 16 by designing and conducting an Anti-Corruption Campaign. They had been working on business ethics subjects since the start of the class, and we were thinking of reaching out to a high school, by visiting the teenagers. Due to COVID-19, we had to design a webinar instead. The college students rapidly designed their project of an Anti-Corruption Campaign for 260 high school students, and we aimed at implementing it on 10 June 2020.

## The Anti-Corruption Campaign

The 33 Ipmi students were organized into eight groups, each of them developing their own unique experiential learning activity. To increase motivation (not that they needed it, but we felt it would add to the adrenaline), we decided the campaigns would be judged by a faculty "jury" who would name the group with the best campaign. Each group used millennials' approaches in the form of apps such as Twibbon, Mentimeter, Kahoot, etc. They created their own games. For example, one of the groups created a "Catch the Corruptor Game" (Figure 16.6). The objective of the game as written by the undergraduate student is as follows: from this game we can learn about that corruption will bring bad impact to everyone including the corruptor. Besides, we can also see from this game, that we need to have a strong commitment to be able to catch the corruptor, and everyone can be a corruptor if they are not diciplined. Therefore, we need to smash bad behavior that will lead us to become a corruptor and we need to work together to end this corruption in Indonesia. This game is really interesting and was inspired by *Mafia*, also known as *Werewolf*, a social deduction game created by Dimitry Davidoff in Russia in 1986.

The big event was conducted as planned on 10 June 2020 and turned out to be a great success. The students prepared their campaign wholeheartedly. The high school students were inspired, which was shown through their commitments which they also posted on social media. At the end of the webinar, the high school students were similarly invited to state their commitments to make change and to post them on social media. This cascading design created positive impacts as students (both the high school and college students) became (a) more aware of the world challenges and the interconnections between the problems; (b) learned about the SDGs; (c) learned about their own areas of control and of influence; (d) gained self-confidence as change makers themselves; (e) promoted the importance of adopting a proactive role for change; and (f) influenced their network of peers in ways we had not considered possible.

At the end of the class, in a separate Zoom meeting with the Ipmi's students, they were asked to express their feelings after conducting the campaign, and they all came up with positive comments (see Figure 16.7).

We asked the college students to rank their learning experience preference, and the first place was given to doing the anti-corruption campaign as a community engagement initiative (see Figure 16.8). As the designers of the anti-corruption campaign, the students also made their own positive commitment to anti-corruption (see Table 16.5).

**FIGURE 16.6** Catch the corruptor game. Source: Amelia Naim's personal documents

**FIGURE 16.6** Continued

**FIGURE 16.6** Continued

FIGURE 16.7 The students' feelings after conducting the campaign. Source: Amelia's Mentimeter Document

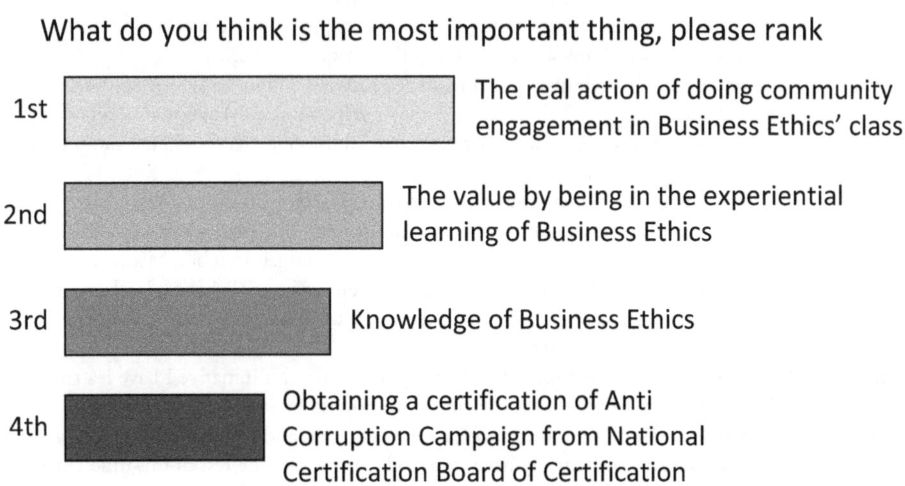


FIGURE 16.8 Rank of preference of class activities. Source: Amelia Naim's mentimeter documents

**TABLE 16.5** Some anti-corruption pledges by IPMI students as the Anti-Corruption Committee

| *IPMI's Students* | |
|---|---|
| Student 17 | Im committed to become a man of anti corruption. I will do my best to prevent it. I will try my best to give lessons for my younger generation. And I will try my best to keep our country clear from corruption. |
| Student 18 | In the future, I will try to keep uphold anti-corruption values, both in my daily life and in my work later. Doing small things like reminding people around not to commit the slightest act of corruption, for example by reprimanding and giving warnings. Through social media like Instagram & Ticktok, I will also convey a positive message about my experience as an agent of an anti-corruption campaign and of course also interesting content related to the spirit of anti-corruption. Even bigger, if there is time & opportunity I can invite my friends to promote anti-corruption campaigns to a wider audience, not only high school kids, but also peers, adults, and even early childhood. I hope this experience can lead me to become a better person and have a high awareness to avoid acts of corruption that can harm others. |
| Student 19 | First of all i will commit in my hearth to not do any corruption practical that bring harmfull for my country. Further, i try my best to share and change the mindset od people starting from my close people then if it any chance to influence more people its my pleasure to do this prestigious things. |
| Student 20 | I would like to grow and create business which is not only beneficial for me but also for the community, even though I am not Indonesia I would like to contribute to the growth of this country to the future and that it is done ethically.Be honest and transparent with all my actions, and fight against corruption. |
| Student 21 | I will remain in the right path, putting justice forward. I want to widen my horizon, and to stay open-minded towards people's perspectives although it contradicts with mine. I will always be against corruption. In addition to that, I will try and put my best for this generation. |
| Student 22 | My commitments are cinta terhadap pekerjaan (Affective Commitment), takut kehilangan pekerjaan (Continuance Commitment), dan adanya rasa kewajiban (Normative Commitment), mempertahankan hubungan tersebut. Misalnya hubungan pacaran, hubungan pernikahan, hubungan persahabatan, dan lain-lain. tidak membuang sampah sembarangan dan bertanggungjawab untuk menjaga lingkungan alam tetap asri. |
| Student 23 | Because of this Business Ethics Anti Corruption campaign Initiative I have lea rned a lot. Everything from what is corruption to how to be integral to my integrity. |
| Student 24 | I will have a commitment to myself for now and then, and I will also keep the commitments that I already have from today onwards. The rest of this time I will use to maximize the realization of my commitments, I will try as hard as possible to maintain my commitment. Strive for my commitment will not betray the results. |
| Student 25 | Be a good, honest man and try to keep fixing the world piece by piece |
| Student 26 | After joining this anti corruption class, I feel more aware to be more commit to myself, to get away from that kind of things related to corruption. And I would like to remain my surrounding as well about how bad a corruption is, we should know between what is wrong and what is right, and after we know, we have to do the right thing, so that we can make Indonesia becoming more a better country. We have to start it from now and even a small things surround us. Always and be brave to remains each other. |

(*Continued*)

**TABLE 16.5** Continued

| *IPMI's Students* | |
|---|---|
| Student 27 | I will apply anti-corruption values in my daily life, and I will commit to not doing corruption in the future because corruption is something that is very detrimental to the country and others, therefore I will serve the nation and this country. |
| Student 28 | I will always remind what my gaols in life, i would never do the corruption, and i always will tell people that they never have to do the corruption |
| Student 29 | I will try my best to avoid doing corruption, and also inspire others to do what is right. encourage others to become an anti corruption people for our self and indonesia |
| Student 30 | Corruptiom is not only corrupting the money. But times and some activities. I will start dicipline myself, and spread awareness to people around me. Because corruption is harming everyone even the corrupter itself. i'm proud to be part of this and this educate myself alot |
| Student 31 | To be a change maker, and to be a person with integrity who will do what's right and not that's easy. I will hold high the justice and do everything I can to keep it. |
| Student 32 | I commit that I will be more responsible with my work and be a good example in the environment, open minded about all things and I will do my best to stand justice for those who needed it |
| Student 34 | I want to use the knowledge that i got to become a better person than before, so that i can make a difference from my sorrounding, and then spread until Indonesia can be a better country |
| Student 35 | I will do my promises to other people so it is fair and respectful it is so that in the future when doing business there will be no corruption |
| Student 36 | I will be become part of of anti corruption. i will tell people to never do corruption and i will do my best |
| Student 37 | I will make a promise to myself that want to make our country better in an anti corruption section and also apply all of ms amel and ms. dewi, ms Julie etc teach to me |
| Student 38 | Im committed from now and in the future I'll do my best to my job, keep it up my application and knowledge of anti-corruption, be honest, not harming others, and try to make this world to a better place |
| Student 39 | I will commit to the nation by preventing corruption because I am very in love with this country, so that this country develops rapidly and becomes a country that can show the best for other countries |
| Student 40 | This is a very exciting, I am increasingly excited to provide counseling to people around me to be able to discipline. I am sure Indonesia can be even better if all can implement anti-corruption, and that all start from ourselves. thank you for the opportunities and that Mrs. Amelia give to us, so on I will make Indonesia better. |
| Student 41 | I want to focus on the importance of integrity. Because integrity is basic things that affects people's behavior and decisions. Especially when they're met with certain situation related with honesty. That's why integrity is so important. |
| Student 42 | I will have a commitment to myself to be honest |

Source: Amelia Naim's document

## Conclusions

We wanted to share this first time experience because as authors we were amazed by its impact and the levels of engagement it created. The learning goals were simple, the design was carefully crafted but did not require major facilitation skills or technical knowledge, other than paying attention to logistical details and operating the platform.

The facilitator designed the session using the Action Reflection Learning principles, a learning methodology originated in Scandinavia in the late 1970s that she codified and has been applying to the design of all kinds of learning environments (Rimanoczy, 2016). This design approach contributed to maximizing the learning and engagement in the first webinar. With respect to the areas of control and influence introduced during the activity, we didn't intend to monitor how far the inspiration, actions and intentions of the initial 100 participants reached, nor did we plan to collect follow-up data about the commitments of the 260 high school children to act against corruption. As educators, we met our baseline learning purpose to introduce the SDGs, teach interconnectedness, promote critical thinking and reflections about values and corruption. We thought that the most important outcome was that we instilled hope and agency in hundreds of young people.

But in hindsight we realize that we went far beyond that. Inspired by the reactions of the initial group of 144 students signing up, and the feedback of the 100 who participated, Amelia took it to the next level. She let herself be guided by the enthusiasm of her students and facilitated the expansion of the activity to a broader audience: 260 teenagers. Furthermore, the Business Ethics students will have an opportunity to apply for their certification as Anti-Corruption Agent-Influencer, undersigned by our National Professional Certification Body. This is possible because they have done all the steps: developing a lesson plan, and arranging role plays, case studies, short essays, and PowerPoint materials to be used in the anti-corruption campaign. They even prepared a program to be implemented offline. So all this evidence will be assessed and they will get their Certificate as an Anti-Corruption Agent-Influencer, which will expand their employment opportunities.

Analyzing what actually happened, we realized that we unleashed the energy and proactivity of all these young people. We are certain that this engaged audience talked about it with their friends and families, shaping a new mindset of empowered action to better our planet. The more we shared this story and the more we pondered it, the more we believed that these young people were now carrying powerful seeds of change. Where could that lead?

As of the writing of this chapter we are exploring different ways to capture and measure the impact of these activities, by reconnecting with the students. But we are also asking ourselves what other possibilities are available to us to develop change makers. Indeed, we heard that the high school students were talking about replicating their experience with elementary school children. What if educators replicated this with their students? What if this became a formal student support for the Office of Anti-Corruption Commissioner in Indonesia? What if this became a student or faculty led initiative to support education and action for the SDGs in different geographies, or influencing decisions of policymakers?

## Limitations and possibilities

We should not end this chapter without mentioning some of the obstacles and limitations encountered. We had some technical constraints, as we depended on good Internet connection for all the participants. We had to limit the first webinar to 100, although some platforms might accommodate a larger number of breakout rooms. The facilitator felt the lack of connection with the audience because they were muted, although she asked everyone to leave their cameras on. This was not possible for some because of a weak Internet signal. In the end, the facilitator asked everyone to unmute their microphones and greet one another and it was a sparkling and chaotic moment that made everyone feel the energy in the room.

As for the possibilities, we haven't finished exploring them. We plan to maintain contact with the participants and send out a survey in six months both to the college and high school students, to hear what they kept or forgot about their initiatives and intentions. This may give us valuable information, although sometimes even what we push aside comes back at a later point, with the force of the important and forgotten things. Amelia will continue teaching this course and Isabel is committed to continue her support. We are aiming at shaping new habits, a new mindset, and expanding consciousness. This will not be done in one day, although every day counts in that journey.

## References

Black, D. (2013). *The Leadership Mandate.* Kindle. Amazon.

Indrajaya, A.N. (2018). Spiritual development programs, individual spirituality and sustainability mindset toward higher commitment to social and environmental impact. *International Journal of Business Studies*, *2*(3), 150–163.

Kassel, K., Rimanoczy, I., & Mitchell, S.F. (2018). A sustainability mindset model for management education. In K.Kassel & I. Rimanoczy (Eds.), *Developing a Sustainability Mindset in Management Education* (pp. 3–37). New York: Routledge.

McCarthy, B. (1990). Using the 4MAT system to bring learning styles to schools. *Educational Leadership*, *48*(2), 31–37.

Rimanoczy, I. (2016). *Stop Teaching: Principles and Practices for Responsible Management Education.* New York: Business Expert Press.

# 17

# REIMAGINING RESPONSIBLE LEADERSHIP THROUGH EXPERIENTIAL LEARNING

## MBA student accounts of engagement with non-profit organizations and their leaders

*Arnold Smit and Armand Bam*

> *I was introduced to a world that is every bit as professional as the for-profit sector, a world that operates on a scientific basis, that thinks deeply about what they are doing and how they are doing it and a world where risk management is inherently part of everything they do as failure has real consequences. If anything, the assignment has certainly broadened my meaning perspective, but it also reignited a lost sense of hope for humanity, that indeed we are able to really care, that altruism does exist and that our skills and capabilities are not solely made for prestige or comfort, but perhaps something more meaningful.*
>
> (Francis Marais, MBA student, University of Stellenbosch Business School)

### Introduction

Despite its natural beauty, its abundance of natural resources, the rich diversity of its people and its vibrant post-apartheid democracy, South Africa is also known for being the most unequal society in the world.[1] While the country is as confronted as the rest of the world with sustainable development imperatives of an environmental nature, it certainly faces a quagmire of social challenges related to poverty, food security, health, education, gender equality and employment.

It is within this challenging context that the University of Stellenbosch Business School (USB) commits itself to responsible leadership development as the definitive goal of all its business and management education programs. Having completed their studies, students should have an ingrained mindset of holding themselves, their teams and their organizations to a higher standard as they serve as custodians of society's institutions and organizations. Apart from offering several other postgraduate programs with similar intentions, the school's MBA aims to develop managers, professionals and entrepreneurs into responsible leaders, expert decision makers, strategic thinkers, effective communicators and stewards of society.

USB's MBA module in Business in Society is especially premised on offering students a transformative learning experience with respect to responsible leadership and societal stewardship. Addressing the role of business in society, the module emphasizes the importance of

DOI: 10.4324/9781003229735-19

personal and organizational leadership that is ethical, responsible and sustainable. The Business in Society module therefore covers foundational themes such as corporate social responsibility and stakeholder theory, followed by practical applications in four domains, namely the workplace, marketplace, environment and community. The module concludes with a focus on ethics, responsibility and sustainability integration with reference to stakeholder reporting, organizational strategy and corporate governance (Cf. Crane et al., 2016).[2]

Students do three assignments during the Business in Society module. All three are premised on building their awareness and reflective capacity regarding responsibility and stewardship in the contexts within which they live and work. One of these assignments is an experiential learning one in which students engage with non-profit organizations (NPOs) and their leaders. This assignment is generally referred to as the MBA Social Engagement Project (SEP). Before we explain more about the assignment itself, a word or two on the theoretical framing of experience and reflection as educational practice.

## Experience, reflection and learning

Business schools are increasingly called upon by students to provide transformative learning experiences as they acknowledge an awareness of their responsibility towards society (Haque et al., 2019). The ability to address the complex challenges faced by societies whether economic, environmental or social needs, requires business schools to move beyond the classroom and present innovative ways of learning. Experiential learning offers this opportunity and can act as an important catalyst in the development of responsible business leaders. Experiential learning's educational intention places students and their actions at the center of learning. This pedagogical approach can allow students to move beyond only talking and observing to developing knowledge and understanding as they "do" and "try" (Dean et al., 2019). In other words, what experiential learning can provide students with, outside of the traditional classroom environment, is the ability to engage with real problems in real time.

Distinct from other forms of learning it leads to deeper levels of reflection, examining beliefs and better meaning-making. Experiential learning and reflection are closely linked, and deep levels of reflection lead to more meaningful learning. While experiential learning has value, the quality of the experience itself must be of a standard that leads students to reflect critically. Inherently, experiential learning hinges on good quality reflection and in return it can improve the capacity to manage complex situations. The ability to think and reflect critically is becoming more valuable for managers of businesses (Bray & Maclellan, 2019). The process of reflection helps leaders to better understand information and determine alternative actions and new ways of thinking. It binds experience and learning through creating time and space for students to step back from what they are doing and consider the meaning of their actions for themselves and others.

Scholars like Mezirow, Dewey, Kolb and others believed that reflection brings about changes in conceptual perspectives. One of the simplest ways of assessing the levels of reflection is to turn to the work of Mezirow and his adult learning theory (Mezirow, 1987). Here student experiences and reflections can be classified as content, process and premise reflection. The most significant of these is premise reflection, more commonly understood as critical reflection. Content and process reflection are shallow reflective states that students engage in daily and may lead to some changes in beliefs but falls short of an internalization of an experience. Critical reflection is the level of reflection students must aspire to achieve. They need to come to the realization of why they do things in the way that they do, why they think the way they do and why they feel the way they do. These understandings are important for their workplace and personal development. Experiential learning should develop opportunities for students to consider more than the "what" and "how" and open them up to understanding why they think and behave in a certain way.

Management education traditionally presents students with opportunities to gain technical skills but the limitations in curriculum design and availability of time for immersive experiences must be corrected. Transformative learning experiences hinge on time for stepping back from an experience and consider all its moving parts; time to deconstruct and reconstruct; and time to make oneself vulnerable to share new learnings. If management educators want students to be equipped for better decision-making, they must lead students to a heightened sense of awareness and social consciousness. The need for learning to be ongoing and active should be cultivated in business schools that are often more concerned with providing students with toolkits to solve generic problems. It is rather in the unconventional ways of student learning that the richer realities can be addressed as part of internships, practicums, group discussions, role plays, self-directed journaling and several other learning experiences.

The ongoing and active nature of reflection should in the end become a way that students embrace their learning not only at a point in time but rather over time. Internalization does not occur unless students are able to do more than observe. Action is required and feelings must be shared and thought processes exposed. It is in many ways an undressing of the consciousness of students that must be laid bare and how they have come to a point in their lives as adults to see the world in the way that they do. This cathartic process can lead to unlearning often not considered as educators develop their teaching materials (Justice et al., 2019). The openness to have one's thoughts and actions critiqued can lead to greater levels of responsibility and readiness for change. New pathways of discovery and knowledge generation must be the aim of experiential learning particularly when students are required to address their assumptions. Leaders are no longer able to live within society without considering the taken for granted social cultural and political domains they function in. They are no longer immune to the impact their actions can have on society because of the businesses they operate in.

Students must be able to move from assessing their experience to analyzing it and ultimately re-imaging ways of seeing the world they operate in. This type of learning opportunity must bring students to a point of clarity and being selective in presenting these opportunities can facilitate this more readily. Placing students at the center of the learning experience will develop their ability to ask meaningful questions and to manage the complexities they encounter daily. The need to critically assess their own assumptions can place them in positions where they come face to face with issues of power. It is in these socio-political experiences that enlightenment may be achieved by reorienting a student's knowledge and belief systems. What is evident is that experiential learning and reflection have value and can lead to changes in the way that students think, act and ultimately lead businesses in society. Included in this leadership role, especially understood from a responsible leadership perspective, is the ability to interact with stakeholders from different sectors. This task includes the ability to lead not only from the center, but also at the margins and especially at the interface between different types of organizations from different sectors of society.

## The social engagement project

The question might well be asked, why MBA students should be sent to learn from NPOs. Our intention with doing so is not about enhancing students' capacity for volunteer work in the social sector. Nor do we intend with this project a form of business school corporate social responsibility through which we might gain reputation mileage for social impact work. The explicit intention is to afford each MBA student that passes through the school, the opportunity to learn about responsible leadership and societal stewardship from people whose daily lives are spent in the trenches of society's sustainable development challenges (Rowe, 2014).

NPOs mainly exist because society does not work in equal measure for the well-being of all citizens. Where governments fail in sufficient and effective service delivery, NPOs will most probably step in to fill the gaps and care for those who are left neglected and marginalized. Where companies externalize social and environmental costs, through the way in which they conduct their operations, NPOs will most likely be at the forefront of advocacy for justice and retribution. The existence of most NPOs are therefore signals of society's fractures in terms of ethics, responsibility and sustainability. This observation is not about claiming impeccability for NPOs, but it at least explains why they mostly originate in the shadow of private and public sector deficiencies.[3]

Sending MBA students, of whom the majority hail from corporate and public sector workplaces, to NPOs, places them in a position of not knowing and opens up an opportunity for learning about leadership and stewardship in complex environments (Bowen et al., 2010). Most importantly, however, our hope is that they will also learn about themselves as they reflect upon their own career and leadership ambitions and the potential of their own positive impact on society and the environment (Mezirow, 1987). This point is aptly illustrated by one student, Esté de Kock (2019) who, at the beginning of the project admitted "when we were presented with the opportunity to engage with and learn from a non-profit organization as part of a project for the module, I thought that we were focusing on the wrong group of stakeholders," only to declare after the project that "this type of authentic leadership style is not just necessary for individuals working in the social sector, but is one that will motivate any team and cultivate an open and transparent corporate culture."

Premised on the potential of transformative learning through reflective and experiential engagement with a NPO, the Social Engagement Project (SEP) assignment is structured in terms of the following guidelines:

- Each student identifies an NPO with a functioning management team and governing body to ensure that the project is done with a legitimate and sizable organization. The NPO's activities should also align with at least one of the 17 Sustainable Development Goals.
- The student visits the organization and studies it within its context to get acquainted with its mission, objectives, operational activities, resource base, stakeholder context and management and governance functions.
- Students are furthermore expected to attend at least a management committee and/or board meeting. The purpose is to immerse themselves in the internal conversations and sense-making discourse about mission achievement amid challenging conditions and resource constraints.

- The experiential engagement concludes with an interview with either the chief executive officer or board chair of the NPO. This is a heart-to-heart conversation in which the student may learn about leadership and stewardship from the NPO leader.
- Finally, the student writes an essay in which they describe what they have learned from the project in relation to responsible leadership and societal stewardship and explain the value that they have gained from it for personal and professional development.

## Student essays

From the more than four hundred students who completed the MBA SEP over a period of two years, we selected four students to write an essay for this chapter. Apart from representing MBA cohorts between 2018 and 2019, they also represent the South African society in terms of gender and racial diversity. The organizations they have done their social engagement project with represent different social mission focus areas, such as employment, education and health.

### *Essay 1: Satisfaction through social engagement – Raven Naidoo*

*When I commenced my MBA journey in 2019, I certainly had a few assumptions and expectations about what the programme would entail. I undoubtedly believed that it would be an intense programme providing knowledge and tools for strategy, efficiency, operational excellence and innovation to pursue the core goal of business, which is arguably driving profit maximisation. I was pleasantly surprized therefore to find that USB's MBA had what I expected, but with a greater focus on the development of responsible leaders that could make contributions to both business and society. The two things that stood out in particular for me was the emphasis on leadership development and the focus on the development of business in society, both which I would later find, had an immense role to play in the total integration of teaching methods and self-immersion of my learning experience which helped me to truly achieve depth in reflection, outward expression and exposure in order to understand society and the world around me whilst completing the programme.*

*The SEP was significant for me in that the projected outcomes of the assignment and the module sought to foster a greater understanding of the need for and importance of aligning the goals of business with the needs and requirements of society and how these factors are not as mutually exclusive as once thought. Whilst not new to NPOs due to prior employment in the sector, the SEP gave me an opportunity to view the sector through a different lens and to immerse myself with the view of an ordinary member of society as opposed to an internal stakeholder. This gave me the opportunity to revisit the thoughts, assumptions and beliefs that I previously had about the world of work and to reframe them in the context of global needs which are exemplified by the advent of the United Nations Sustainable Development Goals.*

*Whilst there are many non-profit and non-governmental organizations established with missions and mandates to provide support to vulnerable and hampered people with social needs in society, there is also a need for mechanisms and organizations that contribute towards sustainable development opportunities to bring about long-term change instead of only providing a short-term solution. It was for this reason that the organization I selected for the SEP was a not-for-profit business incubator, The Furniture Technology Centre Trust*[4]*, which uses entrepreneurship as a tool to empower people to establish a better standard of living for themselves.*

*Undertaking the SEP gave me the opportunity to critically interrogate the methods we apply in business, the outcomes we aim to achieve, the potential beneficiaries of those outcomes and the effect and satisfaction of this on us as individuals. A critical take-home point was that sustainability is the most important aspect of everything that we do today, and this was vital no matter what sector or industry one is involved in. One of the most valuable themes I got out of the SEP was the mindset shift that being a not-for-profit organization should also entail being not-for-loss either. This was pivotal for me as it enhances the view towards sustainability and moves away from the perception of NPOs being dependant solely on philanthropic or state support. This further cemented my learning that there should be no difference between the leadership and management of not-for and for-profit organizations and that both should apply equally the principles of good corporate governance and sound ethical practices underpinned at all times by the drive for sustainability.*

*Undertaking the SEP highlighted the importance of being able to work with a multitude of stakeholders rather than in isolation. The drive for sustainability should be accompanied by a model of joint value extraction and the recognition that business and society are intertwined and therefore require alignment to achieve social and economic prosperity. Further telling for me was the realization that it is not only the collective will of individuals within organizations that are required to achieve its objectives, but also the willingness to share in the implementation and distribution of benefits to society at large.*

*The learning I gained from the SEP was immensely valuable. On a personal level, the project helped to reaffirm my passion and ambitions towards business development as a tool for social upliftment and the development of small businesses. The immersion during the project reminded me that my choice of work can result in far more than just financial remuneration and include the substantial satisfaction of contributing to the development of society as well. It was for this reason that, at the completion of my MBA, I chose to join the public sector in an agency that focuses on small business development and on business development organizations such as the one I selected for my SEP.*

*Undertaking such a project within the MBA programme reaffirmed the challenges that we face as a community, a country, a people and a planet. It has also galvanized the belief that we as individuals and collectives can make a difference if we simply choose to do so.*

## Essay 2: Bringing hope and joy for seriously ill children – Michelle Beukes

*South Africa is a beautiful country with amazing people, yet it is also plagued by severe socio-economic challenges, which include high levels of unemployment, poverty and inequality. In this context, I welcomed the opportunity to participate in an assignment that would allow me to engage with an NPO focusing more on creating social value than financial profit. I recall being excited to learn more about the topics of responsible leadership and societal stewardship.*

*My chosen NPO was the Reach For A Dream Foundation,*[5] *a well-established organization that aims to promote hope and joy in seriously-ill children. The organization achieves this through the fulfilment of "dreams" and facilitate workshops to support the child and their direct family. Reach For A Dream serves the pediatric healthcare sector in South Africa, where there are numerous children, and their families, that are physically, emotionally and financially drained by the diagnosis and consequences of a life-threatening illness.*

*The South African public healthcare sector is widely described as in crisis with understaffed and poorly managed institutions frequently providing inadequate care standards and patient experiences. The private healthcare sector is perceived as offering a higher standard of care, but caters to a minority with out-of-pocket expenses being common, potentially adding substantial financial stress. Sadly, healthcare in South Africa frequently fails to optimally fulfil the care or support requirements of pediatric patients and their families. Reach For A Dream's model provides some response to this public responsibility deficit, in a collaborative and supportive manner that adds value to the communities in which it operates.*

*My decision to engage with Reach For A Dream was based on my professional interest in the healthcare sector and, more personally, as a mother to three beautiful daughters. I wanted to engage with an organization that provides benefits to children. With an average of six dreams being fulfilled each day, Reach For A Dream contributes value by providing positivity and upliftment to those in desperate times, where there often seems to be no hope. The created value also provides a distraction from the unpleasant and support-starved clinical and socio-economic environmental conditions, motivating recipients to cope with far more than just their illness.*

*In the preparation stages for my assignment, I failed to grasp the impact that this project would have on me. It is also clear now that I was particularly naïve in my understanding of an NPO's structure, role and governance, while also being ignorant of the additional pressures and expectations frequently placed on these organizations as they foster a response to an external societal deficit. The privilege to engage with an NPO, impacted me both academically and professionally, as well as personally on an emotional level.*

*Academically, while I found the prescribed material for the module to be interesting, the opportunity to apply the theory to my review of Reach For A Dream brought the literature and its relevance to life. My understanding of the proper governance, and its intention and role in a nonprofit organizational context, deepened when having to assess the principles in reality. Also, being able to view the organization's impact within the construct of its operating environment highlighted the systemic connectivity of various stakeholders and other environmental aspects. Experiencing the course material through real-life engagement, substantially increased my interest and focus on responsible leadership, business sustainability and the rethinking of value creation in business. Wishing to develop my understanding further, I have even decided to choose my MBA electives in alignment with these topics.*

*I learnt that the impact of an NPO can extend well beyond its direct activities and that NPOs form an essential support structure for our country. As for any corporate business, the extent of their reach and the depth of their impact is directly linked to their ability to operate as a well-governed and efficient societal steward. A strategic vision that incorporates principles of ethics, responsibility and sustainability, and has access to adequate resources, is essential to maximize value creation. Through the SEP, I now firmly believe that as individuals, we can all contribute some form of support or value to our environment and those in need. Professionally I have incorporated additional charity support elements into my business and actively strive to develop my competencies as a responsible leader. I have also engaged with Reach For A Dream to investigate the opportunity for me to participate as a dream volunteer in future.*

*Personally, this assignment has had a profound and lasting emotional impact on me. Initially, I was overwhelmed by the sheer volume of sadness and lack of hope that surrounds so many of these seriously ill children. I incorrectly imagined that the only way to cope within such an organization would be to develop a level of desensitization. However, every individual I engaged with displayed a genuine emotional connection to their role and the children being helped. These individuals, some former dream recipients or their parents, all held a personal calling to participate. They are a diverse group of people sharing their passion through purpose, forming the heart of the organization. I shall never underestimate the power of collective, authentic intention again.*

*Finally, on a personal note, every evening, I hug my daughters a little bit tighter, and I am more grateful for their health and our blessings. Participating in the Business in Society module has given me a clearer perspective of the vast contrast in the circumstances of different individuals and the shared responsibility we hold to promote kindness and create value for those less fortunate.*

## Essay 3: Be part of the change – Léander Steynberg

*I decided to work with an NPO in Mbekweni, a community very close, but also starkly different from where I live. Being apprehensive in the beginning, I realized that I can no longer drive by this community every morning and ignore the inequality and pretend that Mbekweni and its problems do not exist. I also knew that becoming involved would force me to have a long and hard look in the mirror regarding my beliefs about and attitude towards NPOs.*

*Mbekweni consists of more than 60 000 people. Unemployment, poverty and other socio-economic and health-related issues, including HIV, are some of the main concerns that have an impact on its wellbeing. It also forms part of a bigger farming community with seasonal workers hailing from the Eastern Cape region in South Africa. The NPO for my SEP was the Be Part Yoluntu Centre (Be Part).*[6] *The organization's vision is to empower the community by creating a sense of awareness about its health and wellbeing. Their credibility is based on their effectiveness in addressing behavioral issues surrounding the spreading and prevention of HIV and their ability to develop new strategies for change. Being situated within a community with a high HIV and tuberculosis (TB) prevalence as well as forming part of a bigger migratory farming community, enables them to penetrate inaccessible areas with different and unique prevention and treatment options to reduce the HIV transmission risk and improve community health.*

*Over the past 11 years, Be Part has built an exceptional rapport with the community. The organization places enormous emphasis on counselling and awareness to address the significant issues relating to HIV, STIs (sexually transmitted infections) and TB as well as wellness interventions aimed at identifying and diagnosing general and specific health risks. This has been a cornerstone of their success in the past.*

*The SEP was for me an exercise in reflection-on-action through which I came to valuable insights and perceptual change. I can now take these changed perceptions and insights gained forward with me when going into a management position as an MBA graduate within a for-profit organization (FPO) to aid others in understanding how NPOs function and operate.*

*I had the perception that NPOs are very different from FPOs and that in general there is less structure in NPOs. Through my conversations with the CEO and Be Part staff, the board meeting that I attended, and the documentation made available to me such as the annual financial statements, this perception of mine was struggling to hold up. I realized that there are more similarities than differences between NPOs and FPOs. I found Be Part to be structured, with functional departments and managers for each. Staff have targets and KPIs and are reprimanded just as is the case at the FPO that I work for. The CEO of Be Part explained that the structure, accuracy and strict protocols demanded by the research work which forms a large part the work at Be Part affects how they approach all their projects. My past engagements with NPO's were very limited which probably contributed to my perception of them being these ad hoc charities where people can go and help out from time to time without any fixed obligations or performance measurements. I thought NPOs were run mostly by volunteers, instead of viewing them as very well-functioning organizations, where volunteering plays a small part in the overall good work that these organizations do.*

*Before the SEP I was under the impression that working for a non-profit is ultimately career limiting. This perception was changed through my conversation with the CEO, a medical doctor, highlighting her personal development as leader of the organization. Amongst other things, she mentioned growing in her understanding of how to run a business as one of the areas in which she developed significantly. She would also not have had this opportunity if she continued with only clinical work in private practice, for example. Further to this, she mentioned the relationships she gets to build and the daily interaction with people on a very personal level as a satisfying experience compared to running a private practice where you would see 50+ patients a day, where there is no time to give anything of yourself, where people become numbers. She mentioned her passion for intensive medicine and research that makes a positive difference in her development. The research allows her to get exposure to testing some of the latest medicine before it goes to the market. She finds her work at the NPO much more stimulating and diverse than being a practitioner. She also feels part of a much bigger international team that is driving global medical research. The balance and diversity between clinical work, business and relationships are what makes her work and development satisfying. This opened my eyes to the fact that at the right NPO a doctor, engineer or a lawyer, or whatever your occupation might be, you can grow and learn just as much or even more than you can if you follow a career in the FPO space.*

*The SEP also changed my perception of the role of NPOs in our society from only being a beneficiary of a company's existing CSR activities, to organizations that can create their revenue from activities that align with their mission and vision. Be Part creates revenue through their research and then, in turn, use the profit from these trials to fund its CSR activities.*

*On a more personal level, this interaction forced me to ask myself whether I am really satisfied with my job at an FPO and whether I feel that it is meaningful. I asked the CEO of Be Part if there are days that her job feels like just a job and nothing more and she was to the point: "Then I will need to find something else to do." She then went on to explain that on days that you do feel negative about the country, as if you do not make any significant change, you need to decide to remain positive, to switch off the negative news coverage and focus on that one person whose life you might have changed instead. Her answer alluded to the fact that for people working within NPOs it is usually about much more than a pay-check. It is about having a purpose and making tangible changes in people's lives. Societal stewardship is described as the responsibility of looking after the social and economic interests of a group of people and this is what they do. NPOs exist due to the failure of the market and/or government. They exist to feed people, to provide healing, and to provide shelter.*

*I have come to the realisation that NPOs take their beliefs and values and put them into action. I found myself almost envying them for the passion and drive they had for the work they do, and I realized that to find purpose in your work within an FPO is perhaps much harder to do than it is in an NPO. I now believe that a fairer, healthier, more equitable world can exist and that NPOs have a major role to play in bringing about the change that is required. I believe that speaking up, getting involved, writing and thinking clearly, and judging situations rationally can also help us to acknowledge the scope of the issues our communities face.*

## Essay 4: Responsible leadership, societal stewardship and motherhood – Elsie Hanyane

*Autism Western Cape's[7] vision is to "optimize the potential of the autistic individual, their families and communities through a comprehensive approach that ensures integration and inclusion in society." Pursuing its mission, the organization focuses on education and awareness, advocacy and equal access, acceptance and inclusivity, empowerment and capacity-building and counselling.*

*I chose the organization because at that time in my journey it had been pivotal in my coping as an African black single mother raising an autistic child far away from any form of family support. I benefited greatly from the organization's assistance which upskilled me to empower others not just emotionally but also in terms of everyday practicalities, for example, providing advice on funding, advocacy and school placement. For me, the espoused and enacted mission and values of the organization were aligned.*

*I was in two minds about engaging with this non-profit for the MBA SEP. My initial thought was that I was familiar with the organization. I anticipated that engaging with them would be straightforward as both the relevant social worker/support group leader and the executive director were familiar with my son and me. We had built a relationship of about three years based on trust and I foresaw that the sharing of information would not be impeded. This would ensure that I had the base I needed to do well in the assignment. On the other hand, I was also mindful about how I felt emotionally about doing an assignment on a topic that was deeply personal and quite uncomfortable at the time. However, from attending the lectures, understanding the objectives of the course and assignment and interacting with the lecturer, I felt that I could trust the process and that sitting with and unpacking the feelings in a safe, facilitated way would have a beneficial outcome for myself and for my parenting style, which would directly affect my son's sense of wellbeing. I realized that in the long term, the advantages far outweighed the discomfort.*

*My view as a responsible leader has shifted. Before this module and the MBA, I was centred on my child and his needs. During my studies, I developed useful skills and insights from various courses on the MBA such as Business in Society, Perspectives on African Frontiers and Digital Enterprise Management. In addition to this, my qualification has augmented my professional network and increased the potential for wealth creation. This means I can have a greater impact on autistic families other than my own. Through doing the Business in Society assignment, I have identified opportunities for autistic children and adults and their families that I can partner with other stakeholders in implementing.*

*As a parent of an autistic son, I previously viewed my ethical responsibility as acting on behalf of my son and less about empowering him. I realize now how my blind spots viewed him more as disabled rather than differently abled. As a result of the reflection during the assignment, my interpretation of ethics is now expanded to how I can start to create an environment, through my influence, that is ready to embrace a differently abled adult in 9 years, so that he can thrive. This is in sharp contrast to my previous views which were how I could constantly plan for how to compensate for everything he is not able to do. The new perspective is exciting and aspirational versus the former which was burdensome and ominous.*

*Leading from the previous point, I now feel that I am empowered regarding my son's autistic abilities. I see opportunity instead of difficulty. My mindset has changed. I am more aware of how partnering with different stakeholders through understanding their needs can create opportunity for both them and the autistic community. I am aware of the first steps required to translate ideas into action, but I am not disillusioned at the effort and commitment required to bring about sustainable change.*

*The SEP also enhanced my understanding of systems thinking. By being more open to listening and putting away preconceived solutions, I got different viewpoints from Autism Western Cape's staff, autistic adults, educators and health professionals that made me realize that many complex factors are at play with any situation. I realize the importance of fully understanding a single action and the ripple effect it can have on various stakeholders and how that in turn can have an added effect on other stakeholders. I realize that it is not about having the perfect solution as much as it is about asking the right questions to the right stakeholders and being willing to share in the journey of achieving shared objectives. Leverage points also become clearer and through an understanding and enquiring mind, I am better able to tap into or influence others to tap into those points more effectively.*

*Although my contact with the NPO has become greatly reduced since the MBA SEP, I continue to search and aim to build sources of community and support around me. I keep hope alive and aim to not only do but be – living an existence that is open to and embracing of neurodiversity.*

## Concluding reflections

Is it possible for students to reimagine responsible leadership through experiential learning? Can such reimagining be leveraged through engagement with non-profit organizations and the constituencies they serve? Furthermore, will such reimagining be sustainable and translate into mindset and behavioral changes with lasting impact on business and society?

As lecturers we dare to believe that transformative and lasting change may result from experiential learning. The four student essays that we have recorded here certainly provide evidence that experiential engagement has the potential to ignite reflection about personhood, responsibility and leadership. Stepping beyond the familiar boundaries of the generally perceived value of an MBA degree for conventional career purposes and engaging with the systemic challenges of their leadership counterparts in non-profit organizations, certainly left these students with a different outlook on their role in both business and society. For them, those parts of society generally served by non-profit organizations are no longer a distant reality, it has become part of one holistic picture, albeit a fractured one characterized by several imbalances and inequalities. Students who embrace and internalize this insight, also realize that responsible leadership entails stewardship of the whole of society and all its stakeholders. The connection between business and the education, health and employment conditions in society – as reflected in the student essays – are no longer accidental, but part of an interdependent whole and within the scope of what the exercise of responsible leadership should be about.

## Notes

1 For more information on South Africa's Gini coefficient and related indicators of inequality visit https://www.imf.org/en/News/Articles/2020/01/29/na012820six-charts-on-south-africas-persistent-and-multi-faceted-inequality.
2 The structuring of the Business in Society module is primarily informed by the framework of Crane, Matten and Spence (2016). Apart from covering fundamental concepts about CSR, the framework also deals with CSR in the community, workplace, marketplace and the environment in separate chapters.
3 Cf. Anheier, 2014, chapters 1–3 for background on the existence of non-profit organizations.
4 https://furntech.org.za/.
5 www.reachforadream.org.za.
6 http://bepart.co.za.dedi189.cpt2.host-h.net/.
7 https://www.autismwesterncape.org.za/.

## References


Anheier, H. (2014). *Nonprofit Organizations: Theory, Management, Policy* (2nd ed.). London: Routledge.

Bowen, F., Newenham-Kahindi, A., & Herremans, I. (2010). When suits meet roots: The antecedents and consequences of community engagement strategy. *Journal of Business Ethics, 95*, 297–318.

Bray, R., & Maclellan, G. (2019). Critical reflection and community education values. *The Journal of Contemporary Community Education Practice Theory, 10*(1), 1–11.

Crane, A., Matten, D., & Spence, L. (Eds.). (2016). *Corporate Social Responsibility: Readings and Cases in a Global Context* (2nd ed.). New York: Routledge.

De Kock, E. (2019). *Student engagement project.* Unpublished essay. University of Stellenbosch Business School.

Dean, B. A., Gibbons, B., & Perkiss, S. (2019). An experiential learning activity for integrating the United Nations Sustainable Development Goals into business education. *Social Business, 8*(4), 387–409. https://doi.org/10.1362/204440818x15445231830058

Haque, A., Fernando, M., & Caputi, P. (2019). The relationship between responsible leadership and organizational commitment and the mediating effect of employee turnover intentions: An empirical study with Australian employees. *Journal of Business Ethics, 156*(3), 759–774. https://doi.org/10.1007/s10551-017-3575-6

Justice, S.B., Bang, A., Lundgren, H., Marsick, V.J., Poell, R.F., & Yorks, L. (2019). Operationalizing reflection in experience-based workplace learning: A hybrid approach. *Human Resource Development International, 23*(1), 66–872. https://doi.org/10.1080/13678868.2019.1621250

Mezirow, J. (1987). Fostering critical reflection in adulthood. A guide to transformative and emancipatory learning. 'How Critical Reflection triggers Transformative Learning'. *Critical Reflection, 74*, 5–12. https://doi.org/10.1002/ace.7401

Rowe, W. (2014). Is nonprofit leadership different from business or government leadership? *Journal of Nonprofit Education and Leadership, 4*(2), 86–91. https://js.sagamorepub.com/jnel/article/view/5934/4777

# INDEX

Page numbers in *italics* represent figures, while page numbers in /bold/ represent tables.

For Product Safety Concerns and Information please contact our EU representative GPSR@taylorandfrancis.com
Taylor & Francis Verlag GmbH, Kaufingerstraße 24, 80331 München, Germany

www.ingramcontent.com/pod-product-compliance
Lightning Source LLC
LaVergne TN
LVHW080809011026
849160LV00015B/1576